STO

**ACPL ITEM
DISCARDED**

Application of Demilitarized Gun and Rocket Propellants in Commercial Explosives

NATO Science Series

A Series presenting the results of scientific meetings supported under the NATO Science Programme.

The Series is published by IOS Press, Amsterdam, and Kluwer Academic Publishers in conjunction with the NATO Scientific Affairs Division

Sub-Series

I. Life and Behavioural Sciences   IOS Press
II. Mathematics, Physics and Chemistry   Kluwer Academic Publishers
III. Computer and Systems Science   IOS Press
IV. Earth and Environmental Sciences   Kluwer Academic Publishers

The NATO Science Series continues the series of books published formerly as the NATO ASI Series.

The NATO Science Programme offers support for collaboration in civil science between scientists of countries of the Euro-Atlantic Partnership Council. The types of scientific meeting generally supported are "Advanced Study Institutes" and "Advanced Research Workshops", and the NATO Science Series collects together the results of these meetings. The meetings are co-organized bij scientists from NATO countries and scientists from NATO's Partner countries – countries of the CIS and Central and Eastern Europe.

Advanced Study Institutes are high-level tutorial courses offering in-depth study of latest advances in a field.
Advanced Research Workshops are expert meetings aimed at critical assessment of a field, and identification of directions for future action.

As a consequence of the restructuring of the NATO Science Programme in 1999, the NATO Science Series was re-organized to the four sub-series noted above. Please consult the following web sites for information on previous volumes published in the Series.

http://www.nato.int/science
http://www.wkap.nl
http://www.iospress.nl
http://www.wtv-books.de/nato-pco.htm

Series II: Mathematical and Physical Chemistry – Vol. 3

Application of Demilitarized Gun and Rocket Propellants in Commercial Explosives

edited by

Oldrich Machacek

Universal Tech Corporation,
Dallas, Texas, U.S.A.

Kluwer Academic Publishers

Dordrecht / Boston / London

Published in cooperation with NATO Scientific Affairs Division

Proceedings of the NATO Advanced Research Workshop on
Application of Demilitarized Gun and Rocket Propellants in Commercial Explosives
Krasnoarmeisk, Russia
18–21 October 1999

A C.I.P. Catalogue record for this book is available from the Library of Congress.

ISBN 0-7923-6697-2

Published by Kluwer Academic Publishers,
P.O. Box 17, 3300 AA Dordrecht, The Netherlands.

Sold and distributed in North, Central and South America
by Kluwer Academic Publishers,
101 Philip Drive, Norwell, MA 02061, U.S.A.

In all other countries, sold and distributed
by Kluwer Academic Publishers,
P.O. Box 322, 3300 AH Dordrecht, The Netherlands.

Printed on acid-free paper

All Rights Reserved
© 2000 Kluwer Academic Publishers
No part of the material protected by this copyright notice may be reproduced or utilized in any form or by any means, electronic or mechanical, including photocopying, recording or by any information storage and retrieval system, without written permission from the copyright owner.

Printed in the Netherlands.

TABLE OF CONTENTS

Preface	vii
Acknowledgements	ix
Introduction, *B. Matseevitsch*	xi
Opening Remarks, *O. Machacek*	xiii
Class 1.3 Composite Propellants as Ingredients in Commercial Explosives – The UteC Experience, *O. Machacek, G. Eck and K. Tallent*	1
Commercial Explosives on the Base of Removed Energetic Condensed Systems (Conception, Problems, Decisions), *B. Matseevich*	7
Demilitarization of Large Rocket Motors and Propellant Utilization, *W.O.Munson*	17
Utilisation of Double-based Powders and Rocket Propellants for Production of "Dry Explosives of Slurry Type", *B.Vetlický and T.Dosoudil*	25
Industrial Explosives from Recovered Powders and Solid Rocket Propellants, *E.F.Zhegrov and E.V.Bercovskaya*	29
Application of Cryocycling to Rocket Motor Propellant Size Reduction and Reuse, *J. Lipkin, L.R. Whinnery, S. Griffiths, R. Nilson, J. Kaminska, G. Mower, W. Munson, J. McNair and J. Elliott*	35
Cofiring of Propellant Washout Residue with Traditional Boiler Fuels: Resolution of Operational and Environmental Issues, *S. G. Buckley, J. Lipkin, L. L. Baxter, R. Moehrle, J. R. Ross, G. Mower and W Munson*	37
Conversion of Demilitarized Explosives and Propellants to Higher Value Products, *A.R. Mitchell, M.D. Coburn, R.D. Schmidt, P.F. Pagoria and G.S. Lee*	49
Some Aspects of the Application of Small Grain Powders in the Emulsion Explosives, *P.Kohlicek, E. Jakubcek and S. Zeman*	59
Joint Demilitarization Integration, *J.Q. Wheeler and J. Lipkin*	73
New Aspects of the Development of the Industrial Explosives and Materials on the Basis of Utilized Powders, *N.G. Ibragimov, E.F. Okhrimenko, E.H. Afiatullov, U.M Yukov, U.M. Ivanov and I.H. Garaev*	79
Small-sized Sources of Seismic Signals on the Basis of High Energy Condensed Systems for Transition Areas Ground – Water and almost Inaccessible Regions of a Land, *N.M. Pelykh, N.M. Pivkin, A.P .Talalaev , R.P. Savelov and Y.A. Byakov*	85

Physical and Chemical Principles of Slurries, *C.-O. Leiber and R.M. Doherty*	91
Detonation Properties of Gun Propellants and Slurry Explosives and their Combination, *H. Schubert*	111
Reproductive Technologies of Gunpowder Production, *F.F.Raran*	115
The Use of Surplus Smokeless Powder Propellants As Ingredients In Commercial Explosive Products in the United States, *G. Eck, O. Machacek and K. Tallent*	119
Burning and Detonation of Water-impregnated Compounds Containing Solid and Liquid Propellants, *B.N. Kondrikov, V.E. Annikov and V.Yu. Egorshev*	133
Characterization of Intermolecular Explosives, *R.M. Doherty and C.O. Leiber*	141
Safety Aspects of Slurry Explosives, *Nico H.A. van Ham*	149
Use of Converted High Energy Value Explosive Materials as Industrial Energetic Materials, *N.K.Saygin, B.V.Matseevich, V.P.Glinski, O.F.Mardasov and N.I.Plechanov*	175
The Technology of Slurry as Dispersing Media for Powders and Propellants, *J.C. Libouton*	181
Promising Boosters for Blasting Commercial Explosives in Boreholes, *V.P.Iliyn, A.G.Gorokhovtsev, Y.S.Kulakevitch, N.I.Rabotinsky and C.P. Smirnov*	189
The Application of Reclaimed Explosives in Commercial Emulsion Explosives, *N.I. Rabotinsky, V.A. Sosnin and V.S. Iliukhin*	193
Utilization of composite propellants with obtaining specific products, *A.J.Salko, A.P. Denisjuk, Yu.G. Shepelev and A.B. Vorozhtsov*	199
Creation of Safe on Manipulation Industrial Explosives and Products for Mining Industry on the Basis of Gunpowder V.P.Glinskiy, *O.F.Mardasov, N.V.Mochova, N.K.Shaligyn and B.F.Obrazcova*	203
Basic Directions of Works on Utilisation of Pyrotechnic Products and Ammunition, *N.M.Varenych and V.G. Dzhangarian*	209
Powerful Insensitive Hybrid Explosives using Inorganic Propellants/ Pyrotechnics in Conjunction with Organic CHO Compounds for Tailorable Blast Applications, *A.J.Tulis*	215
Conditions of Ammunition Utilisation in People´s Republic of China, *Tsin Chanjung*	233

PREFACE

This book contains papers presented at the NATO Advanced Research Workshop titled "Application of Gun and Rocket Propellants in Commercial Explosives". (SST.ARW975981)

The workshop was organized in collaboration with codirector Dr. Bronislav V. Matseevich (KNIIM) and held in Krasnoarmeisk, Moscow Region, Russia, October 18-21, 1999.

About 70 participants from 11 different countries took part in the meeting (Russia, Belarus, Czech Republic, Germany, Belgium, China, USA, Spain, Israel, Ukraine and the Netherlands).

The workshop was principally the continuation of a previous NATO workshop on Conversion Concepts for Commercial Application and Disposal Technologies of Energetic Systems" held at Moscow, Russia, May 17-19, 1994 in the specific area of the reuse of gun and rocket propellants as ingredients in commercial explosives.

Oldrich Machacek

ACKNOWLEDGMENTS

I would like to thank Dr. B.V. Matseevich, Director of the Krasnoarmeisk Scientific Research Institute of Mechanization ("KNIIM") for his extensive involvement as co-director in organizing the Advanced Research Workshop in Krasnoarmeisk, Russia.

Special thanks goes to Dr. V.P. Glinskij, Dr. I.V. Vasiljeva and A.I. Fedonina from KNIIM and Dr. B. Vetlicky for invaluable assistance in preparation and the smooth operation of the workshop.

I would like to express our gratitude to the Advisory Panel on Disarmament Technologies of the NATO Science Advisory Committee for granting the financial support for the meeting especially Prof. H. Schubert and Mrs. Nancy Schulte in instigating and encouraging the workshop. Last but not least, our thanks go to all the authors and participants for their presentations, contributions and dedication to this successful meeting.

Oldrich Machacek

Introduction

B. Matseevich

Dear participators of work conference!

Let me to welcome all of you on Submoscow land.
Leading specialists, prominent scientists, engineers from many countries of the world have gathered within these walls to discuss one of the most important direction in up-to-date engineering - rational application of removed energetic materials such as powders, propellants and explosives in interests of all people progress.
"War arsenals to transfer to service of peace". Without exaggeration this brief statement is a motto of our conference in which participators from twelve countries such as USA, Germany, Belgium, Netherlands, Peoples Republic of China, Israel, Canada, Spain, Russia, Ukraine, Byelorussia, Czech Republic take part.
Geography of cities from which representatives have arranged is spread over all clock zones of the world. It reports about deep interest of many countries, firms in exchanging of experience about this important question. There are more than 150 methods of utilising explosives, powders and propellants usage in civil industries interests; and only explosion energy usage in interests of mining, metallurgical machine building industries can absorb to 95-98% of all released materials.
That's why in the name and themes of our conference this problem appoints as the general one.
Discussed problem demands of solution for wide variation of tasks: scientific, technical, economical and organising. The workers of mining industry which will use these materials naturally will ask the questions, will danger of blasting works increase when military explosives, powders and propellants are introduced in compositions of commercial explosives. What economical, technical and ecological advantages can made favourable attitude of these workers to new commercial explosives, new manuals for these explosives use in competition with traditional compositions.
To give answers to these questions, which are universal for all countries of the world, scientists have to work together. Pooling of experience is of most importance.
For this purpose during last four years Krasnoarmeisk Institute of Mechanisation had organised three conferences, two of which were Russian and one was international with countries of SNG. In every conference to 250 specialists from more than 90 organisations have taken part. Common opinion is exclusive benefit of these contacts, and published transactions are the textbooks for specialist study. That's why we give great importance to our work conference

that unites scientists of many world countries, suppose these contacts are very useful and necessary.

There are few days left before 2000 year comes, this is one of cardinal landmark in people history. Let our united work to serve one of steps forward to progress, maintenance and securing peace and understanding between peoples of all countries in the world.

On behalf of our organising committee let to wish success in work, health and happiness to all of conference participators and their families, organisations, firms, countries.

Opening Remarks

O. Machacek

Let me make some remarks to the opening of this Advanced Research Workshop concerning the use of military propellants and explosives as ingredients in commercial explosives. The idea for this workshop started when Dr. Schubert visited UTeC's facility or what we call the "PRUF" Plant, located in Hallowell, Kansas. This facility is using shredded propellants from large rocket motors as an energetic ingredient in packaged explosives, which are used in coal mining and quarries. You will hear more about the UTeC facility in the next presentation.

Dr. Schubert mentioned that the NATO Scientific Division has a strong interest in demilitarization activities and will support any technical and scientific information exchange between scientists from NATO and CIS countries. Scientists from other countries are also welcome to participate.

The idea of using the excess military explosives and propellants in commercial explosives is obviously not new. With the end of each big war, there was always a tendency to use the leftover explosives for something useful. The closest thing was always the commercial blasting industry. In the United States after WWII and the Korean War, large amounts of smokeless powders were used for blasting by the Hercules Corporation, which is now a part of DYNO NOBEL.

With the recent end of the Cold War, there is a realization on the part of the military, that there are too many explosives and propellants in storage magazines. Why do we need such a large quantity of small arms ammunition that we are capable of killing the entire population of the world several times over? The development of modern, more sophisticated weapons made such an approach even more irrational.

A long time age, one of my teachers used to say "There has to be a war every twenty to twenty-five years because the stability of smokeless powders is twenty to twenty-five years and the military has to get rid of the stuff". There was some rational in this approach.

The 20-century, which is now coming to close, has been the bloodiest century in the history of mankind. We do not have to remind ourselves about WWI, WWII, Korean Vietnam Wars and other conflicts in different parts of the world. There is reasonable hop to believe that the 21-century will be different. There is the realization that the solution to some differences by war is not necessarily the optimum solution. The world has become too small for that. But that's enough of philosophy.

The idea to use the military's excess explosives as ingredients in commercial explosives seems to be attractive. Why not use the energy, which was purposely put into these materials, to do something useful, like mining or breaking rock, etc.? There are some technical, regulatory, applications, economic and other problems that we will be talking about. As the graph shows, vast amounts of commercial explosives are used in the United States. Similar, probably slightly fewer amounts are used in Russia, China, India, S. Africa, Australia, etc. These amounts are staggering. Most people do not realize that in the United States alone, there are more commercial explosives consumed in one month than the total amount of explosives used in WWII. However, there is one significant difference, commercial explosives are inexpensive compared to military explosive. Practically all-modern commercial explosives are based on Ammonium Nitrate, which is also used as a fertilizer. The introduction of military type explosives and propellant into commercial explosives presents a significant economic challenge. Sometimes, even if you get the military propellants for nothing, you still have a hard time competing on an economic basis with regular commercial explosives. Many of the reasons include more complicated regulations, stricter handling practices, etc. I think this workshop will address some of these problems.

We also would like to emphasize, that the purpose of this workshop is not only to present and to listen to presentations of some nice papers, but also encourage working relationships between individuals and organizations from all sides in order to find out if we can do something together in the future. I hope we will have enough time to accomplish our goals in and out of the sessions, meetings and discussions. There is a lot of explosive talent and experience gathered in this room. So let's try to be successful.

Now let me introduce the first speaker.

FIGURE 1
SALES FOR CONSUMPTION OF U.S. INDUSTRIAL EXPLOSIVES

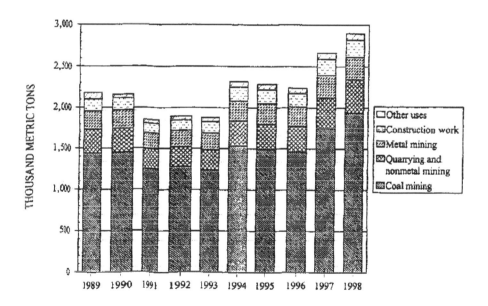

CLASS 1.3 COMPOSITE PROPELLANTS AS INGREDIENTS IN COMMERCIAL EXPLOSIVES - THE UTEC EXPERIENCE

O. MACHACEK, G.ECK, K.TALLENT
Universal Tech Corporation, Dallas, Texas

1. Introduction:

This paper presents information concerning UTeC's findings on the use of rocket propellants as ingredients in commercial watergel explosives. The composite propellants discussed were derived from either demilitarization efforts or manufacturing excess produced by many of the Defense and Aerospace Industries. The paper covers some of UTeC's historical involvement in the recycling of composite propellants into commercial explosives from one-kilogram lab samples to a successful 20,000 lb. (9,1 kg) per day of commercial explosives production facility.

2. History:

UTeC received its initial exposure to the opportunities in the recycling of Large Rocket Motor propellants through a contact with CSD in early 1992. CSD and UTeC entered a joint venture investigating the development of a commercial explosive formulation containing a Class 1.3 composite propellant from the Minuteman III Stage III Rocket Motor. This propellant was a typical Ammonium Perchlorate (AP)Aluminium (Al)/Rubber Binder type propellant. During 1992, UTeC was successful in developing a large diameter packaged commercial explosive product containing up to 30% composite propellant.

Following the success of the initial work, CSD and UTeC continued their efforts with the full-scale demonstration of the initial work done in Phase I. During this time, a full-scale demonstration of the manufacturing and testing of a 15,000 lb. (6,8 kg) batch of a packaged watergel slurry blasting agent was accomplished. At the same time, the product formulas were tested and the US Department of Transportation awarded the formal shipping approvals.

3. The Process:

The manufacture of watergel slurries incorporates a basic two step process. First, a mother solution, or liquid phase (which normally constitutes 30-60% of the final product) is produced. The second step constitutes the blending of the mother solution with any additional dry nitrate salts, aluminium powders (if necessary) and gelling agents. The blending process is utilized to homogenize the mixture and to entrain air into the mixture. Air is entrained until the density of the product is achieved. When ready to package, the mixture has the consistency of thick oatmeal, which is easily pumped into packages. The necessary equipment consists of temperature controlled storage tanks, a mixing chamber and packaging equipment.

In order to incorporate the composite propellant, an additional step must be added. This step was the downsizing of the propellant into a useful and manageable dimension. This was found to be approximately 1-inch (25-mm) pieces. The method chosen to accomplish this task was a counter-rotating disk shredder. For safety reasons, the shredding process was conducted under a liquid (mother solution) and was conducted remotely. This added step would prove to significantly slow down the production rate over the standard 0% propellant slurry. For this reason, production costs were somewhat higher than those encountered for the production of most large diameter blasting agent explosives. Therefore, in order to make the final composite containing blasting agent economically competitive, it has often times been necessary to charge a processing fee for downsizing the propellant into an acceptable form for use as an explosive ingredient.

4. Performance Testing:

During the initial work, several explosive product formulations were tested for their performance characteristics. Many of the early formulations proved not to work at all. These formulations were quickly abandoned, while efforts increased with the more promising formulations. The most promising formulations were based on a Hexamine /Ammonium Nitrate watergel slurry matrix. The shredded composite propellant was added to the standard Hexamine/Ammonium Nitrate slurry in different weight percentages. These test mixes were 100 standard slurry + 0% propellant, 90% standard slurry + 10% propellant, 80% standard slurry + 20% propellant and 60% standard slurry + 40% propellant. The three experimental products and the standard slurry were tested on a comparative basis. The two main test matrices were the Velocity of Detonation (VOD) Test and the Underwater Energy Test (Bjarnholt). The test results were as follows:

TABLE 1: Unconfined VOD Test results for Class 1.3 composite propellant mixes at 22°C (meters/second)

Mix Descriptions	Charge Diameter (inches/mm)		
	4/102	3/76	2/51
Standard Hexamine Slurry	4,600	3,900	Fail
90% STD + 10% propellant	4,060	3,860	3,210
80% STD + 20% propellant	3,950	3,540	3,080
60% STD + 40% propellant	3,790	3,500	2,830

TABLE 2: Unconfined VOD Test results for Class 1.3 composite propellant mixes at 5°C (meters/second)

Mix Descriptions	Charge Diameter (inches/mm)		
	4/102	3/76	2/51
Standard Hexamine Slurry	4,010	3,450	Fail
90% STD + 10% propellant	3,910	3,410	Fail
80% STD + 20% propellant	3,680	3,130	Fail
60% STD + 40% propellant	3,400	2,870	Fail

TABLE 3. Underwater Energy Test Data for Class 1.3 composite propellant mixes (cal/gm)

Mix Descriptions	Shock Energy	Bubble Energy	Total Energy
Standard Hexamine Slurry	373	414	787
90% STD + 10% propellant	370	434	804
80% STD + 20% propellant	398	470	868
60% STD + 40% propellant	449	525	974

NOTE: The initial testing was conducted using the Minuteman III Stage III propellant, however, subsequent-testing on similar propellants has yielded similar results.

A product formulation utilizing 20% of the shredded propellants was selected to go into the product of a commercial blasting agent. It was felt that this amount was optimal considering the test results, the anticipated production rates and the slurry's final rheology for mixing and pumpability. Currently UTeC uses a formulation containing 23% shredded propellant.

5. The PRUF Plant (Part I):

As a result of UTeC's successes, UTeC in cooperation with CSD, designed a facility dedicated to the processing of demilled or excess Class 1.3 rocket motor propellants and their subsequent use as an energetic ingredient into a commercial explosive product. This facility located in Southeast Kansas become known as the PRUF Plant (Propellant Re-Use Facility). The PRUF Plant housed a propellant shredder as well as the final product mixing and packaging equipment. The PRUF Plant construction was completed in the fall of 1995.

In the PRUF Plant production process, approximately 1,500 lbs. (680 kg) of propellant (usually in a block configuration weighing no more than 18 kg each) were placed onto a belt conveyor. The propellant was then remotely fed into the awaiting solution filled shredder. The shredder downsized the propellant into irregularly shaped pieces with an average maximum cross section of 3/n-inch (19-mm). The wet propellant was collected in a flooded hopper beneath the shredder. Once the shredding was completed, a gate valve on the bottom of the hopper was opened and the propellant/solution mixture was dumped into a transfer tank. This tank was transported to a position above the slurry mixer, where the propellant was added to the mixer through an overhead chute. When the appropriate amount of propellant had been shredded and transferred to the mixer, the other ingredients (Ammonium Nitrate, additional Mother Solution, Guar Gum, etc.) were added, and the final Class 1.5 blasting agent slurry was made. The slurry was then packaged into 4-inch (120-mm) to 6-inch (152-mm) diameter shot bags.

Almost immediately after production began, a design flaw was discovered in the shredding operation. The wet shredded propellant would stick together or bridge in the flooded catch hopper located below the shredder. This unexpected situation resulted in many lost production hours, as personnel had to often times manually dig the propellant out of the hopper after each shredding operation. Even with the slower than expected production process, UTeC manage to process over 350,000 lbs. (159 kg) of over 10 types of Class 1.3 composite propellants. This resulted in the production of over 1.5 millions lbs. (681,82 kg) of finished explosive product between the Fall of 1995 and the Fall of 1996.

6. Election Day - The Set Back:

In November of 1996, a major set back in the operation of UTeC's PRUF Plant was encountered. During the shredding process, a video camera above the shredder became obscured by the splashing solution. Because of the poor visibility, the operator could not see that the level of solution was low and that 7, four-pound (1.8 kg) cylindrical shaped motors had not been pulled through the shredder, but were rolling on top of the blades. The shredding operation was completed without incident. However, during the cleaning of the shredder blades after the shredding process was completed (rotating the shredder blades in reverse and then forward), one of the propellant motors was pulled down into the shredder blades and began to be shredded. Without the presence of the surrounding mother solution, the propellant motor caught fire and quickly ignited the other motors. Additional motors waiting to be shredded also caught fire as the fire spread down the conveyor's belt. As a result of the fire, the Pruf Plant sustained approximately $500,000 damage and was shut down for all of 1997. Luckily there were no injuries and no explosions.

7. Lessons Learned:

1. Wet propellant bridges (don't' let the propellant collect)
2. Round propellant casings roll (don't use round or cylindical configuration)
3. Don't store propellant nearby when shredding
4. Don't store any unnecessary combustibles nearby when shredding
5. Use a low liquid sensor in shredder to prevent running the shredder dry
6. Use some type of fire suppression system over the shredder and conveyor to control propellant fires
7. Improve the camera optics to provide a better view of the shredding process
8. Update SOP's to include what to do when the camera is obscured or broke

8. PRUF Plant (part II):

It was quickly decided to rebuild a better PRUF Plant. In order to get a more efficient production rate; the propellant hopper/gate valve assembly was abandoned in favour of a system that would continually take away the shredded propellant. This would eliminate the propellant bridging problem. In the new process the shredded propellant is continuously taken away from the shredder by an auger system and is collected in a hopper. The shredding process is repeated three times or until approximately 3,500 lbs. (1,590 kg) of shredded propellant has been collected. The mixing and packaging process of the final blasting agent was left unchanged.

In addition to the new auger system, a couple of safety systems were also added. These included a low liquid level sensor in the shredder housing, which would not allow the shredder to run if the mother solution level was below the shredder blades. The other new system was a fire suppression system (deluge system) directed at the shredder, the conveyor belt, and the auger. The deluge system was designed to keep any potential fire started during the propellant shredding process from spreading to other equipment.

Since the re-startup in March 1998, the PRUF Plant has processed approximately 1.5 million Lbs. (681,82 kg) of Class 1.3 composite propellants and produced approximately 6.5 million lbs. (2,954.550 kg) of finished Class 1.5 blasting agent explosives. The current production rate is approximately 4,500 Lbs. (2,050 kg) of propellant processed per day, or approximately 19,500 lbs. (8,860 kg) of finished blasting agent per day. Some of the propellants that have been processed through the PRUF Plant are from the following projects: Titan, Orbus, Minuteman and Hawk. UTeC is currently searching for additional sources of these types of propellants, which may be of use in the PRUF Plant process.

9. Summary:

UTeC has shown that the use of composite propellants as energy enhancing additives in commercial explosives is a viable method for their use. The application of this alternate use technique has positive implications for the beneficial use of excess materials from the composite propellant inventory and it makes a very good commercial explosive product. It should be noted that each material must be evaluated for this process and not all the propellant formulations are accepted. The PRUF Plant manufactured blasting agent, Slurran 600-20, has performed very well and has been well received in the Southern US market. UTeC and United Technologies have received two US Patents on this technology (US Patent No. 5,536,897 and US Patent No. 5,612,507).

COMMERCIAL EXPLOSIVES ON THE BASE OF REMOVED ENERGETIC CONDENSED SYSTEMS (CONCEPTION, PROBLEMS, DECISIONS)

B. MATSEEVICH
FSUE"KNIIM" Krasnoarmeisk, Moscow region, Russia

Methods of conventional ammunition demilitarization by use of detonation, open burning, sinking in the depth of world ocean, which up-to-date existed in our country and abroad, had been refused by world society as they caused great damage to environment and resulted in irretrievable losses of valuable for mankind materials contained in ammunition.

So conception of ammunition utilization have been developed in Russia as a general approach to realizing the problem of demilitarization ammunition written oft because their storage period is finished or to taking off obsolete systems of weapon or liquidating weaponry in according to international disarmament treaties. This conception is built on few general principles:

1. Comprehensive conversion of ammunition and ammunition components.

Utilizing process has to provide conversion of all ammunition elements including warheads, powder charges and motors, initiating systems, control systems, packing and so on.

2. Safety of utilization processing.

Utilization process in most cases is more hazardous than loading process as for some objective causes (wide variation of ammunition constructions concentrated in single production, various conditions of ammunition storage and exploitation, difficulties of disassembling and removing explosive and so on) as for subjective difficulties caused by slight researching of unloading processes, little experience of our country utilization industry, organizing problems of ammunition delivery to demilitarization and so on.

So special complex of methods (technologies and specified equipment) has to be designed depending upon type of explosive, powder or propellant, dimensions, weight, construction of ammunition; and also there have to be dissolved problems of controlled ammunition delivery to utilization, production projecting and exploitation, technological discipline and personnel training.

3. Utilization processes have to be environmental acceptable.

When ammunition is burned on open sites or exploded large quantity of toxic oxides, cyanides, salts of heavy metals dioxins is got into environment. Air, water and soil have become dirty. So utilization technologies have to be provided for prevention of environmental damage.

Using utilization processes have to be realized with minimum economical losses, and when deep secondary conversion of received row materials is carried out it has to be economically profitable to the exclusion of individual ammunition classes and types processing.

There are many types of ammunition classification which are defined either usage scheme or belonging to certain kind of military forces or military materiel. These classifications solve problems of army or designers, but they do not suite to utilization production organization.

Ammunition classification due to its adaptation to utilization is based on following general principles: ammunition ascribes to one class in according to unity of technological solution and processes for realizing the most hazardous operations of unloading.

The choice of suitable utilization process depends upon ammunition construction, type of explosive and necessity of removed explosive preparation to next conversion.

Due to these conditions utilized ammunition is recommended to divide into six classes. It is obviously before ammunition utilization starts a stage of ammunition disassembling is realized and then a stage of explosive removal from ammunition case is carried out. The practice shows the most dangerous operation is explosive removal from ammunition case.

Recently a few technological processes and equipment are developed that provides to select for every ammunition class or group of classes the most economical and suitable for organization and technical reasons design decision of utilization production organization due to classification mentioned above (Fig.1).

Technology and equipment for explosive removal from ammunition have to connect with next directions of removed explosive, powder and propellant conversion and usage in country economy. Design of products on the base of utilized ammunition explosive

- main customers of designed explosives are mining enterprises, geology and others

- commercial explosives on the base of explosives and powders (energetic materials - EM) must not be less technological and environmental safety in comparison with wide used commercial explosives

- technological processes of products manufacture from EM must be adapted if possible to existing productions ammunition loading and common commercial explosives productions;

- production of designing explosives may be realized not only on plants, but on bases and arsenals where ammunition is stored;
- design explosives must be delivered to customer for manufacture of blasting works in order to requirements of Gosgortechnadzor of Russia.

It has to decrease much the amount of designing commercial explosives because every new composition requires blasting works operators adoption to it.

On the base of this approach and analysis of composition-technological features all removed EM were divided into four main groups. This classification of EM enables to design rational set of commercial explosives from removed explosives and powders of each group, to provide them for correspondent standard documentation and to realize method of their production on plants.

Ammunition division into classes, explosive removing methods, division EM into groups and methods of commercial explosive production and types of commercial explosive are schematically represented on fig.2-6.

More detailed classification is represented on fig.4.

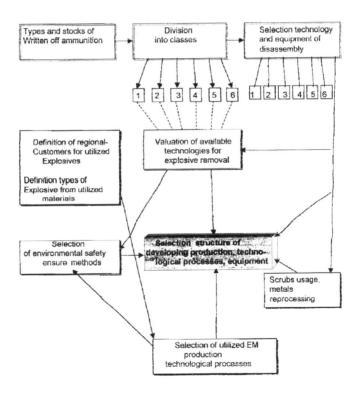

Figure 1. Block scheme for structure of comprehensive conventional ammunition utilization production design

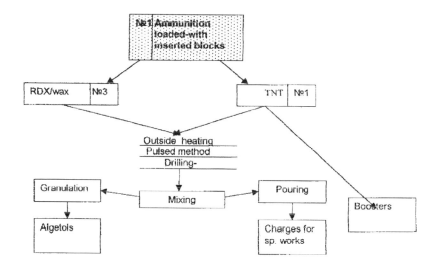

Figure 2

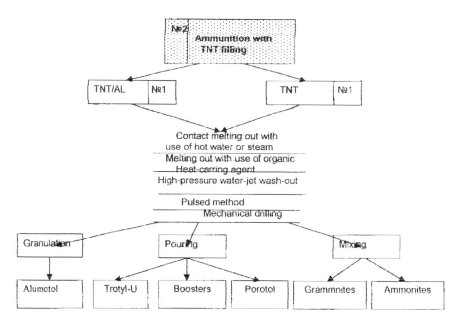

Figure 3

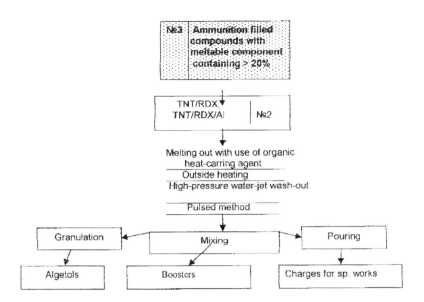

Figure 4

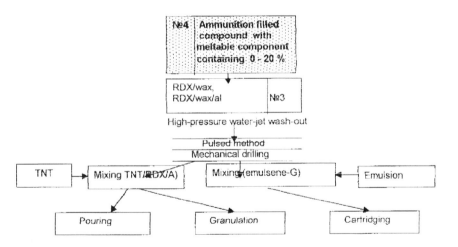

Figure 5

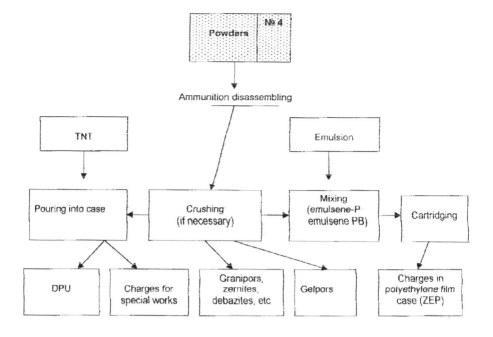

Figure 6

TABLE 1. Commercial explosives designed with use of explosive component from utilizing ammunition

Group	Name of utilizing EM in group (amount of EM utilization to 2000 y	Name and brief characteristics of obtained explosive	Standards for design products and method of their production
1	2	3	4
1	Trotyl and its mixture (TNT, Al, etc.) (70 000 t)	1.1. TROTYL-U. It's intended for charging boreholes any degree of watering to blasting works manufacture in any climatic zone. It's produced as pieces of various forms and dimensions with maximum size of piece 45 mm.	TU 7511809-80-93 Mechanical crushing of utilizing charges. (Designer KNIIM)

Continuation of TABLE 1.

1	2	3	4
		1.2. GRAMMONITES 30(10, 40160. They are intended to produce blasting works on earth surface for borehole of any degree of watering. They are produced from Trotyl-U or any Trotyl makes as granules hemispherical form with maximum diameter 8 mm.	TU 7511819-91-94 Dry granulation. (Designer KNIIM)
		1.3. Boosters T-400G. They are used as boosters for initiating hole charges of low-sensitive commercial explosive on earth surface. They are produced as a mixture of Trotyl-U TU 7511809-80-93 and Trotyl make A GOST 4117-78 or Trotyl GOST 7059-73.	OCT 84-411-80 Pressing. (Designers GOSNII "Cristall" and KNIIM)
		1.4. SEISMIC WAVE SOURCES I S-100. They are intended for initiating resilient ocsillations in earth crust by energy of explosion to producing seismic works with use boreholes of any degree .of watery. They are produced as charges-in polyethylene case Trotyl-U filled.	TU 7511809-78-92 Pressing. Assembling (Designer KNIIM)
		1.5. ALUMOTOL. It is intended to produce blasting works on earth surface for borehole of any degree of watering. It is produced from utilizing composition TNT/Al as granules semispherical form with maximum diameter 8 mm.	GOST 1269C-77 Dry granulation. (Designers GOSNII "Cristall" and KNII)
2	Melting compositions contained RDX, TNT/RDX, TNT/RDX/Al/ wax type (20 000 t.)	2.1 ALGETOLS They are intended to produce blasting works on the earth surface in boreholes of any degree of watery. They are produced from utilizing explosive TNT/RDX, TNT/RDX/Al/wax type. They are the most powerful explosives from existed one. It's designed 3 makes: Algetol-15, Algetol-25, Algetol 35	TU 7511819-89-94 Dry granulation (Designer KNIIM)
		2.2. SEISMIC WAVE SOURCES IS-500, IS-1000. They are intended for initiating resilient oscillations in earth crust by energy of explosion to producing seismic works with use boreholes of any degree .of watery. They are produced as charges in polyethylene case TNT/RDX/Al/wax type <u>composition filled</u>	TU 7511809-78-92 Pouring. Assembling. (Designer KNIIM)
3	Dry or moisture composition contained RDX: RDX/wax RDX/wax/Al, (7 000 t.)	3.1. ALGETOLS	See gr. 2

Continuation of Table 1.

1	2	3	4
		3.2 EMULSION EXPLOSIVES (Emulsene-G) with diameter 36, 60, 90, 120 mm. They are intended to produce blasting works on earth surface in boreholes and cavities of any degree of watery.	TU 07511819-109-98 Mixing emulsion with RDX containing compound. (Designer KNIIM)
		3.3 SEISMIC WAVE SOURCES IS-100 They are intended for initiating resilient oscillations in earth crust by energy of explosion to producing seismic works with use cavities and boreholes of any degree .of watery. They are produced as charges in polyethylene case RDX/wax and RDX/wax/Al types compositions filled.	TU 7511809-78-92 Pressing. Assembling (Designer KNIIM)
4	Powders and propellents (130 000 t.)	4.1 EMULSENE -P. It's represented oil-in-water emulsion of ammonium and sodium nitrates water solution sensitized by powder.	TU 7511819-98-95 Mixing emulsion with powder Designer KNIIM)
		4.2 EMULSION-POWDER CHARGES They are intended for manual charging on the earth surface. They are produced from Emulsene-P as cartridges with diameter 45, 60, 90 and 120 mm, waterproof.	TU 7511819-90-94 Cartridging. (Designer KNIIM)
		4.3 GELPORS They are represented water contained slurry on the base of water oxidizer solution sensitized by powder or propellent crumb. They are produced as cartridges for blasting works on the earth surface for dry or watery boreholes.. They are produced as cartridges with diameter 90 or 120 mm.	TU84-75090-66-93 Cartridging Designer FGUP "SOUZ" with KNIM)
		4.4 POROTOL. It is intended for loading seismic, crushing, flat, combined and other articles to producing special blasting works. It's represented a composition of pyroxyline powder grains and trotyl, waterproof.	TU 7511809—83-93 Pouring. (Designer KNIIM)
		4.5. SEISMIC CHARGE ZS-701-3. ~ It's intended for blasting works in boreholes filled with water or solution. They are produced as charges in polyethylene case filled with porotol.	TU 4112-080-91 Pouring (Designer "VNIP Ivzrivgeophizika" with KNIIM
		4.6. CRUSHING FLAT COMBINED CHARGES (ZDPK). It is intended for secondary crushing big pieces of rock. It's represented a casting butt-end of which is steel covered. It's in 1.5 time more effective than analogy trotyl charge.	ZDPK-OOOTU Pouring, Assembling (Designer KNIIM)

Continuation of Table 1.

1	2	3	4
		4.7 MODULAR COMBINED CHARGES (ZKM) It's intended to producing blasting works on the earth surface for reclamation canal building and cleaning, crushing rock, outline explosion. It's produced as a block of propellent charge and Trotyl	TU 7511809-83-93 Pouring, Assembling (Designer KNIIM)
		4.8. ZERNITES. They are intended to produce blasting works on the earth surface in boreholes of any degree of watery. They are produced as pouring mixture on the base of grain powder, water and oxidizer solution.	TU 84401-181-93 (Designer CNIIHM with KNIIM)
		4.9. GRANIPOR PPF. It's intended to produce blasting works on the earth surface in boreholes of any degree of watery. It's represented mixture of grain powder and fuel oil, waterproof.	TU 0751819-96-95 Mixing (Designer KNIIM)
		4.10. BOREHOLE COMBINED CHARGES. They are intended for blasting works on the earth surface. They are produced by hole filling of interchanged lays of ammonium nitrate and I grain powder.	TU 0751819-95-94 (Designer KNIIM)

Certain characteristics of designed explosives are represented in table 2.

TABLE 2. Characteristics of designed explosives

Explosive name	Characteristics						
	Explosion heat, kJ/kg	Flash point, °C	Sensitivity to impact, GOST 4545, device No.1	Sensitivity to friction, OST 84-895, MPa	Density, Kg/m^3	Detonation velocity, km/s	Oxygen balance, %
Trotyl-U	3900	295-305	0-20	>500	750-800	5,0-5,0	-74
Grammonite 30/70	3768	315-320	12-24	216-284	800-900	3,8-4,5	-45,9
Grammonite 40/60	3747	320-325	12-24	216-284	800-900	3,7-4,4	-36,5
Algetol-15	4735	210	8	294	900-1000	4,6	-80,8
Algetol-25	4860	210	8	294	900-1000	4,8	-78,8
Algetol-35	4986	210	16	294	900-1000	5	-75,9
Emulsene-G	4291	230-240	0-8	350	1450-1480	5,4-6,0	-16,0
Emulsene-P	3200	190	0	>300	1500	5,2-5,6	-15,6
Gelpor – 1	3771	170-185	0	220-250	1300-1400	5,0-5,2	-0,8…-14,0
Gelpor - 2	4190	175-185	0	220-250	1300-1400	4,5-5,3	+3,0…-8,0
Gelpor - 3	3561	170-190	0	157-220	1300-1400	5,1-5,3	+1,9…-6,7
Porotol	3875	170-180	52	108,9	1500	6,5	-59,4
Granipor PPF	3436	180-190	8-12	200-205	800-900	5,5-6,3	-42…-45

The efficiency of explosive may be valued on criterion specific power which is included base explosion characteristics: p, kg/m^3 ; explosion heat Q, kJ/kg, detonation velocity D, m/s.

Designed explosive efficiency data are represented in Table 3.

TABLE 3. Efficiency of commercial explosives on the base of utilizing EM

Explosive	Characteristics			
	P, kg/m^3	Q, kJ/kg	D, m/s	$N_{sp.}$, $(kW/m^2)10^9$
1	2	3	4	5
Trotyl-U	750-800	3900	5500	17,16
Grammonite 30/70	800-900	3768	4500	15,26
Grammonite 40/60	800-900	3747	4400	14,84
Algetol-15	900-1000	4735	4600	21,78
Algetol-25	900-1000	4860	4800	23,33
Algetol-35	900-1000	4986	5000	24,93
Emulsene-G	1450-1480	4291	5700	36,08
Emulsene-P	1500	3200	5600	26,88
Gelpor-1	1300-1400	3771	5200	27,45
Gelpor-2	1300-1400	4190	5300	26,42
Gelpor-3	1300-1400	3561~	5300	27,15
Porotol	1500	3875	6500	37,78
Granipor PPF	800-900	3436	6300	19,~18
Ammonite 6GW	800-850	4315	4000	14,67

From the data of tables 2, 3 it is followed that all design commercial explosives on the base of utilized EM exceed ammonite 6GW in specific power. Sensitivity of design granulate explosives to mechanical influences is on a level with amnonite 6GW.

The most safety to mechanical and heat influences and the most environmental acceptable are emulsion explosives.

Laboratory and industrial tests of design explosives on quarries of AO "Transvzrivprom", AO "Gidrospecstroi", OAO "Karelski okatish", OAU "Kovdorski GOK" confirm their technical and economical efficiency, which is provided not only lower cost in comparison with common explosives, but higher energetic characteristics of new explosives.

Realization of designs on plants and mining enterprises ensures:

1 Development of scientific-technical base of utilization, that is completely explosive removal, utilizing EM preparation and conversion in production for blasting works in industry.

2 Design and delivery to customer new effective commercial explosives, part of which had been designed for the first time and has no analogies in Russia and abroad.

DEMILITARIZATION OF LARGE ROCKET MOTORS AND PROPELLANT UTILIZATION

W. O. MUNSON
Thiokol Technologies International Inc.
P.O. Box 707
Brigham City, Utah 84302-0707

1. Introduction

Interest in demilitarization activities was intensified by three significant events in the early 90s: 1) the reunification of Germany which, accompanied by the withdrawal of the Former Soviet Union's (FSU's) military, left enormous amounts of weapons to be destroyed; 2) the establishment of a continuing line-item budget in the United States (US) to dispose of its obsolete and over-aged munitions; and 3) US congressional funding directed to provide technologies for large solid rocket motor (SRM) disposition. The focus in Germany was to destroy the munitions with minimal impact to the environment. Thus, the selected technology was incineration with accompanying gas clean-up systems. In the US, the stated goal was to maximize resource recovery and reuse (R^3) of the demilitarized items. This goal was traded-off with the costs associated with disposal efforts. The result for munitions was a balance among four options: R^3, confined burning/detonation, controlled destruction (plasma, molten salt, etc.) and open burning/open detonation. The large SRM disposition effort was funded through the Joint Ordnance Commanders Group (JOCG) SRM subgroup. Initial focus was on R^3 methods.

Thiokol played a major role in developing SRM demilitarization activities. Our focus was to improve our existing SRM demilitarization process through waste-stream minimization and find a use for the removed propellant. Through internal and JOCG funding, we developed a closed-loop process for treating washout water. This process entailed removing the water-soluble materials from the water by an evaporation recrystallization process. Remaining water was returned to the washout process. The second activity was to find a use for the removed propellant. Two approaches were taken: 1) reutilization of SRM propellant in commercial explosives or blasting agents (the subject of this paper), and 2) recovery of propellant ingredients such as ammonium perchlorate (AP). Recovery of AP was identified for Thiokol's hazard class (HC) 1.3 propellants, since there is a market for the recovered material. However, this process leaves an AP-depleted residue (mainly hydrocarbon binder and powdered aluminum) that has not, as yet, found a beneficial use. Thiokol also focused on determining the best method to reuse the propellant in commercial explosives or blasting agents.

2. Propellant Removal and Size Reduction

Thiokol Propulsion has been demilitarizing large (i.e., over 2 metric tons (MTs)) SRMs for over 33 years. Methodologies employed include water washout (hydromining), dry machining, and burning in the open atmosphere. The method used most frequently is hydromining. Over 90 percent of the propellant removed from SRMs at Thiokol (equaling over 12,000 MTs of propellant) has been by this method. We have removed propellant from approximately 4,500 motors with 30 different types of configurations. The first hydromining facility was constructed to recover metal cases for reloading propellant. More recently, the focus has been to minimize open burning of the propellant and recover the major propellant constituent AP.

The largest motors hydromined have been Space Shuttle SRM segments which are 3.7 m diameter, 8.7 m long, and contain 127 MTs of propellant. A special facility (Figure 1) was built in 1998 to accommodate these large segments. By the end of 2000, a total of 16 segments will have been hydromined, equaling over 2,000 MTs of propellant.

Figure 1. Space Shuttle 200 MT segment in washout stand

The largest quantity of propellant hydromined from a given SRM family is 8,000 MTs from approximately 350 first-stage Minuteman missiles. Figure 2 shows a first-stage Minuteman in our other hydromining facility.

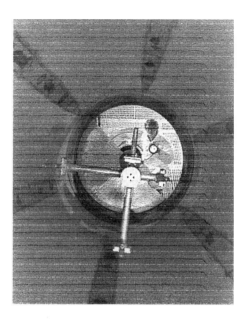

Figure 2. Demilitarization of a Stage 1 Minuteman

Thiokol's process begins with a loaded motor case (LMC) that is shipped to the hydromining facility. The LMC is transferred to the washout station on a wheeled dolly using a rail-to-rail transfer approach. The transfer is powered by a winch system located in the facility. Forward and aft shrouds are attached to the LMC in the facility to contain the water and washed-out propellant. The aft shroud is designed to accept the washout boom and direct water and propellant coming from the motor into the separation and recirculation system. The LMC is secured to the washout stand, and the LMC and stand are tilted to a 30-degree angle with hydraulic cylinders.

Hydromining is accomplished with the boom traveling slowly up the LMC center bore. The boom is rotated slowly as it moves into the LMC. High-pressure water (~680 atmospheres) is directed onto the propellant through a series of nozzles. The nozzles are arranged to cut-out chunks of propellant. The propellant and washout water flow by gravity out the aft shroud and onto a vibrating screen that separates the liquids and solids. The boom-nozzle arrangement is adjustable, in that the nozzles can maintain a 5-10 cm distance from the propellant surface as the propellant is cut away. Hydromining is controlled remotely from a bunker. Adjustments can be made while the system is shut down and after viewing the results of previous cuts.

Washout water from the screen is further filtered to remove solids and then pumped back to a holding tank where it is recirculated into the high-pressure pumps. Since some of the propellant ingredients are soluble in water, they could build-up in the recirculation loop. This build-up is monitored and controlled either by a slip-stream (sent to another part of the process), by cooling the water and recrystallizing some of the soluble ingredients, or by using both approaches.

The wet solids from the screen are transferred to size-reducing equipment that produces propellant chips less than 1.5 cm in major dimension. If the propellant is to be used in a blasting agent, size reduction takes place in a desensitization fluid composed of water and soluble salts, such as ammonium nitrate and sodium nitrate that are used in the final commercial blasting agents. The solution-to-propellant ratio is kept above 10:1 to assure process safety through thermal control and wetting propellant surfaces. Material from the size-reduction step is separated into wet-desensitized propellant (approximately 60 percent propellant) and desensitization fluid. The fluid is returned to the size-reduction system surge tanks. Make-up material is added to the tanks to account for the fluid lost with the propellant. The desensitized propellant is loaded into reusable drums and transported to a packaged-product facility or a mine site where it is mixed with bulk blasting agent and loaded directly into bore holes.

The propellant-removal and size-reduction system is designed to eliminate production of any hazardous waste products. All of the propellant, water, and desensitizing solution are used in blasting agents including washdown water from external and internal equipment cleaning.

3. Blasting Agent Development and Utilization

Surplus military explosives and small arms and artillery propellants have been used in the mining industry for decades. In the US, this utilization was especially prevalent after World War II with the introduction of slurry blasting agents into the marketplace. Surplus explosives and propellants were often used as sensitizers and fillers. Essentially, all the propellants used were single-, double-, or triple-based powders. Little or no composite AP-based rocket propellants were used. As mentioned above, in the early 90s attention was focused on disposition of large strategic SRMs to meet age-out and treaty requirements. At that time, there was also an emphasis on waste minimization and R^3 of demilitarized weapons.

In the US, the JOCG SRM subgroup sponsored work on safe, environmentally sound, and cost-effective disposition of large SRMs. One of the projects dealt with demonstrating the viability of using SRM propellants in commercial blasting agents. Two contracts were awarded by the JOCG: one to a team of Thiokol Propulsion and Dyno Nobel Inc., and the other to a team of United Technology (Chemical Systems Division) and UTeC Corporation. Both teams successfully demonstrated that SRM propellants could effectively and efficiently used in commercial mining. The following summarizes the Thiokol-Dyno Nobel program.

Our formulation and testing protocol followed a series of tests and evaluations. Initial work was done using a theoretical performance code to determine the effect of propellant type and percentage on key explosive parameters such as detonation velocity and pressure, oxygen balance, major detonation products, and initial explosive density. In parallel, a series of small compatibility tests were conducted to assure the various ingredients used in propellants and those used in blasting agents were stable when mixed together. Tests included differential scanning calorimetery, differential thermal analysis, and accelerated rate calorimetery. After assuring compatibility, a series of small mixes were made to determine basic safety properties. These tests included responses to impact, friction, and electrostatic discharge, and determining an auto-ignition temperature. After establishing that safety properties were in the acceptable processing range, kilogram-sized mixes were made. These mixes utilized levels of propellant from 10 to 40 percent to determine processing properties (rheology, working life, etc.) and performance properties (minimum booster, detonation velocity, critical diameter). Based on the test results, the level of propellant addition was selected at 30 ±5 percent. Larger batches of the selected formulations were made (100-200 kg) for United Nations (UN) transportation tests and accelerated aging tests. After passing the UN HC 1.5 testing and demonstrating good stability on aging, several thousand kilograms were made of four formulations. Two water-gel formulations were made: one containing AP-based, HC 1.3 propellant and the other containing nitroglycerine HMX HC 1.1 propellant. Two emulsion formulations were made: one with HC 1.3 propellant and the other with HC 1.1 propellant. These blasting agents were tested in both hard-rock mines and coal mines. The propellant-containing blasting agents performed equal to or better than the ammonium nitrate fuel oil (ANFO) and heavy ANFO currently in use. Typical data for the propellant containing blasting agents are shown in Table 1.

Thiokol has discussed reutilization approach with FSU experts. A concern expressed by potential Russian users was that the propellant-containing blasting agent would produce hydrogen chloride (HCl) in the detonation products. Their propellants contain up to 70 percent AP. Previous Russian mining experience using neat AP as an explosive was disappointing because the HCl produced caused worker reentry problems. However, we demonstrated that by including a stoichiometric amount of sodium or calcium nitrate in the mixture, virtually no HCl is formed as shown in Table 2, which is based on theoretical equilibrium calculations. Subsequent closed-bomb testing with kilogram samples at the New Mexico Institute of Mining and Technology verified the absence of HCl in the detonation products.

In a preliminary evaluation of disposition of Russian strategic SRMs, Thiokol sponsored work under the direction of the Russian JSC Technical Chemistry at the Research and Development Institute of Polymer Materials (NIIPM, Kirov Plant), Perm Russia. Kirov made blasting agents containing Russian propellants. Preliminary results (Table 3) showed that effective blasting agents can be made with the Russian SRM propellants.

TABLE 1. Comparison of propellant-containing blasting products with standard blasting products

Theoretical	Emulsion (EM)		Water Gel (WGM)	
Percent HC 1.1	25	---	36	---
Percent HC 1.3	---	25	---	20
Gas Volume (moles/kg)	35	35.6	35	37.7
Oxygen Balance (percent)	-6.06	-4.35	-2.79	-1.48
Detonation Products (mol/kg)				
Al_2O_3	0.88	0.74	1.27	0.59
$CaCl_2$	0.12	0.74	---	---
NaCl	---	---	0.34	1.18
$CaCO_3$	0.72	0.1	---	---
$NaCO_3$	---	---	0.38	0.04
SiO_2	0.24	0.24	---	---
H_2O	23.87	25.18	23.67	25.1
CO_2	2.93	3.43	3.47	4.04
C	1.89	1.36	0.87	0.46
N_2	8.16	6.99	7.90	8.39
HCl	---	---	---	---

TABLE 2. Russian Tests (NIIPM, Kirov Plant)

Characteristics	Brands (Base)	
	Aquapan (gel)	Emulpan (emulsion)
Estimated explosion heat, kcal/kg	880	920
Charge density, kg/m^3	1,200	1,300
Estimated energy concentration, kcal/dm^3	1,056	1,196
Critical detonation diameter, mm	65	45
Rate of detonation, km/sec		
• In various experiments	3.8-4.2	4.2-5.2
• After storage in real-life conditions	4.0 3 months at t=0 to +20°C	4.7 1 month at t=+5 to +20°C
Crater-forming TNT equivalent α	0.7	0.72
Hazard class under UN regulations	1.5	1.5

TABLE 3. Performance of propellant containing blasting products

	Propellant Products			
	Emulsion (EM)		Water Gel (WGM)	
Percent HC 1.1	25	---	36	---
Percent HC 1.3	---	25	---	20
Measured				
Density (g/cc)	1.32	1.33	1.49	1.20
Detonation Velocity (m/s, 150 mm)	6,270	4,890	6,490	4,430
Total Energy (cal/g)	951	776	1,033	845
Relative Properties				
Energy (cal/g), % Theoretical	0.94	0.81	0.95	0.87
Relative Weight Strength, Experimental, ANFO = 1	1.08	0.88	1.17	0.96
Relative Bulk Strength Experimental, ANFO = 1	1.62	1.33	1.98	1.31

Based on our efforts in the US and FSU, our conclusion is that with the demonstrated and robust technology available today, serious consideration should be given to removal and commercial reuse of SRM propellants in commercial blasting agents.

UTILISATION OF DOUBLE-BASED POWDERS AND ROCKET PROPELLANTS FOR PRODUCTION OF "DRY EXPLOSIVES OF SLURRY TYPE"

VETLICKÝ B., DOSOUDIL T.
Pardubice, Czech Republic

1. Specific questions of use of disassembled and inveterate double-based smokeless powders and solid propellants

In consequence with the tradition and with the unification within of framework of former Warsaw pact and with a wide assortment of powders in comparatively small amounts, came in our country into being a comparatively complicated situation. Fundamentally, without any problem was the utilisation of small-grained powders for infantry and mortar ammunition – already many years ago was introduced an industrial explosive PERMONEX BP, being a mixture of smokeless small-grained powder with ammonium nitrate for compensation of oxygen balance. This explosive was loose and by execution analogous to ammonites.

More complicated was the problem of double-based tubular powders and rocket for artillery ammunition and rockets. The complexity reposed on the fact, that as a liquid nitroester was in our country used both nitro-glycerine and diglycoldinitrate, for most of calibre both alternatives, being fully equivalent from the ballistic point of view (strategic reasons – raw materials basis). In the case of rocket tubular elements was for the rocket 130 mm JRRO used a powder on the basis of diglycoldinitrate without addition of heavy metals compounds as a burning modifier. At the beginning, tubular powders in filling bundles were dropped into bores with a bigger diameter and fired. The detonation was mostly not full and sometimes with a secondary flame in the mixture with the air too. However, in more soft rocks a positive effect was achieved.

Under these circumstances we get down to the development and investigative work with a target to ensure better utilisation of dismantled powders and solid propellants for rockets as industrial energetic materials. Our sight was also supported by the fact, that the production capacity of spherical powders was not used.

2. Philosophy of the resolution

Double-based smokeless powders, being considered for the utilisation, have energetic behaviour more or less close to TNT (for various calibre dimensions they vary between approximately 700 kcal/kg, (2930 kJ/kg) to about 900 kcal/kg (3768 kJ/kg). A typical presentation of explosive changes is burning. For achievement of better behaviour from the point of view of detonation possibility we utilised the principle of spherical powders manufacturing for preparation of a porous product. This can be realised on the equipment for making spherical powders by means of increase of temperature in the course of solvent removal at agitating the mixture in reactor. Porous powders in dry state highly incline to detonation and the oxygen balance was compensated close to zero by means of addition of granulated ammonitrate, according to the powder composition.

3. Experimental part

As a basic raw material we used tubular rocket powder based on diglycoldinitrate. It was necessary to disintegrate them mechanically in advance, because the dissolving in original state would last unproportional long time. As in the case of spherical powders, ethyl acetate was used as a solvent. Prepared granulation product was to be dried originally, from the manipulation safety point of view we tried to prepare a granulation product in the mixture with ammonium nitrate, i.e. a certain analogy of trinitrotoluene slurry of the 1st generation.

Granulated product contained after processing at the suction nutsch filter about 32 – 35 % water, in the mixture with nearly triplicate amount of granulated AN the total water content reduced to about 12 % The mixture had a loose character and was practically flame-resistant, because a substantial part of water was bonded in pores of the powder. At the beginning, we investigated this energetic material with very strong initiating charge (SEMTEX). Later on we decreased the primer and in the end we established, that for the initiation is completely sufficient a detonator N. 8.

4. Experimental results

DETERMINATION OF SHOCK SENSIBILITY

The test was carried out according to the proposition of the revision of the Standard ČSN 66 8071, conformable with the Recommendation of UNO (Recommendation on the Transport of dangerous Goods, Tests and Criteria, UN, New York, 1990, Test 3 (a) (ii)).

Testing conditions
Conglomerated particles of the test specimen should be crushed and screened, for testing should be used the portion passed through the sieve with a sieve mesh of 0,5 mm.
Dosed: 40 mm^3

Mass of the hammer: 10 kg
Activation evidence: explosive decomposition

Powder mass RM
Results:
Falling height 50 cm: 6 x not any reaction
The specimen can be not activated nor at use of maximum incidence energy, being available at the testing equipment.

Mixture of powder mass RM + AN (explosive)
Results:
Falling height 50 cm: 3 x activation, 3 x not any reaction
Falling height 40 cm: 6 x not any reaction
Minimum incidence energy $E_{dmin} = 50$ J

Determination of friction sensibility
The test was carried out according to the Recommendation of UNO (Recommendation on the Transport of dangerous Goods, Tests and Criteria, UN, New York, 1990, Test (b) (i)). and in accordance with ON 66 8095.

Powder mass RM
Results:
At the pin load 36 kg: 6 x not any reaction
The specimen can not be activated nor at maximum prescribed load.

Mixture of powder mass RM + AN
Results:
At the pin load 36 kg: 6 x not any reaction
The specimen can not be activated nor at maximum prescribed load.
Note:
Owing to possibilities of testing equipment was the friction sensibility test carried out at higher loads with following results:

Powder mass RM
Results:
At the pin load 38,5 kg: 3 x smoke 3 x not any reaction

Mixture of powder mass RM + AN
Results:
At the pin load 38,5 kg: 6 x not any reaction
At the pin load 43,2 kg: 2 x smoke, 4 x not any reaction

Development of smoke in the course of testing indicates a decomposition of tested specimen but not its explosive change. By this reason, obtained values do not have a relation to the friction sensitivity of the material.

TABLE 1.

Order N.	Test type according to ČSN	Testing conditions	Results	Note
1	Detonation velocity	t = 10 °C b = 98, 66 kPa atmospheric humidity = 60 %	2625 m/s	ČSN 66 8066/ A/9b
2	Performance by Trauzl test – ballistic mortar		88,4 %	ČSN 66 8075
3	Brisance by Hess		14,7 mm	ČSN 66 8065
4	Detonation ability by detonator Cu N. 3		not detonating	
5	Detonation ability by detonator Cu N. 8		satisfied	ČSN 66 8070/B
6	Performance by Trauzl test		320 ml	ČSN 66 8064

5. Conclusions:

By means of described technology it is possible to prepare an industrial explosive, safe from the manipulation point of view. The explosive is loose and reliably detonates by the aid of standard detonator N. 8. For preparation can be used common equipment for manufacture of spherical powders after a change of the technological regime. From the point of view of performance and outputs it could be classified as a common industrial energetic material.

INDUSTRIAL EXPLOSIVES FROM RECOVERED POWDERS AND SOLID ROCKET PROPELLANTS

E.F.ZHEGROV, E.V.BERCOVSKAYA
FCDT"«Soyuz", Dzerzhinsky, Moscow Region, Russia

Powders and solid rocket propellants (P and SRP), phased out and disarmored from state reserve, have the rather miscellaneous nomenclature both on chemical composition of propellants, and on charges used in artillery and in rocket motors (about 70 chemical compositions and 300 types of charges). P and SPR differ by a rather broad spectrum of physicochemical properties, explosive performances and geometrical sizes (Fig 1). By sizes a range of variations of values is from a fraction of mm up to meters. By chemical composition - from a practically pure nitrocellulose up to high filled compositions containing metallic fuel, explosives (nitroglycerine, hexogen, dazin and others), and also the whole gamma of different liquid and powdery inert components. From a point of view of operation explosive performances P and SRP, on the one hand, considerably differ depending on a chemical composition, and, with another, all of them essentially rank below nominal explosives, having major sensitivity to mechanical effects and smaller to a shock wave (Tab. 1.).

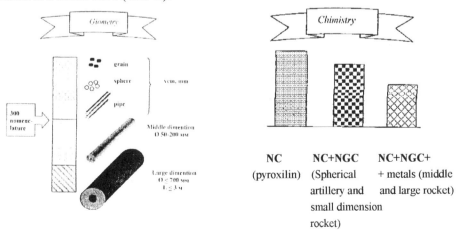

Fig.1 Classification of pand SPR by sizes of charges and kinds of chemical composition

TABLE 1 Explosive performances P and SPP

The mark	A critical diameter of a detonation, $d_{cr.}$, mm	Critical pressure of shock in detonation, $P_{cr.}$, kbar	Sensitivity to abrasion at strike shear, kgf/cm^2	Sensitivity impact, % detonatings
RST	1,8 – 2,5	40	1089	DBSRP
RSI	3 – 4	67	1270	50
NMF-2D	18 – 19	80	1573	55
PP		40		
Trotyl	**8 – 10 (c = 1,0)**	**19,6**	**>3000**	**4-8**

In this connection the problem of usage P and SRP in civil purposes develops into a severe engineering problem, which one is aggravated by that usage of powders on straight assignment (in a combustion regime) in mass quantity is insignificant (Fig. 2).

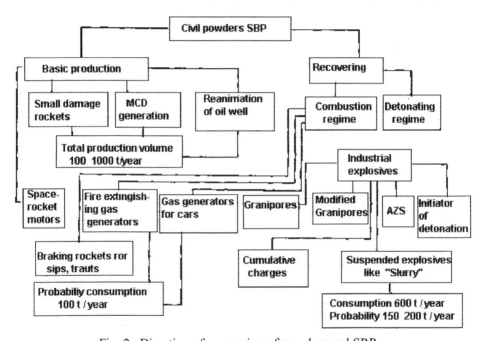

Fig. 2. Direction of recovering of powders and SRP

The most expedient direction of the utilization P and SPR, having in view of the market of consumption, is usage them as civil explosives.

However ensuring of indispensable production characteristics of these explosives requires (demands) to solution of three problems:
- Lowering of sensitivity to mechanical effects;
- Sensitization to a shock wave;

- Improving of sanitary-hygienic parameters ecology both commodity and detonation products.

In Tab. 1 the comparative data of powders and explosive on sensitivity to mechanical effects and shock wave and in a Fig. 3 - ways of improving of these performances are presented.

The sensitivity to exterior abrasion and impact is reduced by covering of granulouse powders with surface-active agents (in particular by mineral oils), and powder grains - film of inert materials (cardboard, pergamine and other). Sensitivity to a shock wave is increased by special initiators put as the cartridge into a turned jacks or pasted on a back of an article as discs in thickness 5 ... 10 mm. The material of these pioneers represents a thermoplastic composition with a major content of hexogen or another explosive.

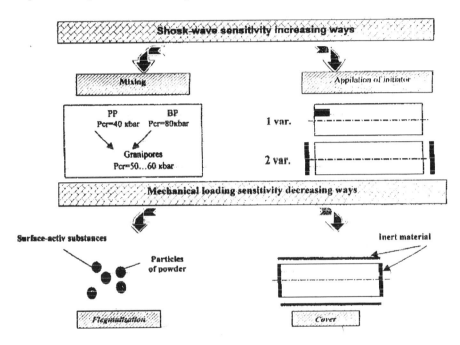

Fig. 3.

The sanitary-hygienic problems at usage P and SRP are defined by two factors: by a major content in a composition of nitroglycerine (28-40 % at $LPC_{without\ with} = 0,02$ mg/m^3) and toxicity detonation products (WITH and NO). The lowering of toxicity is realized at the expense of covering of charges inert materials and raising of an oxygen balance of explosive composition by usage of caliche.

On the basis of engineering solutions, presented above, a series of explosives covering practically all the nomenclature of recovered P and SRP (Fig. 4) is developed. Certificated and found in industrial service: granipores, DSC, shaped charges. The industrial trials of monolithic blasthole charges and gelatinous water containing compositions ("gelpores") are conducted. All the developments have patents for the inventions.

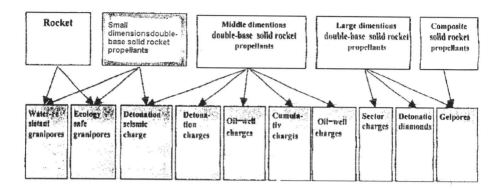

Fig. 4. Nomenclature of explosives based on powders and propellants

Among the merits of all explosives on the basis P and SRP it is necessary to refer them practically unrestricted water resistance and long terms of operation operability. It is impossible to eliminate and state relevance of a problem of rational usage of military commodity, in a great quantity deposited in arsenals of different regions of the country.

Technology processes of manufacture of all nomenclatures of explosives, basing on equipment, P and SRP are designed.

From a point of view of execution distinguish three groups of explosive exacting definite architecture of production process:

- Granulouse explosives such as "Granipores", received by grinding of artillery pyroxiline and ballistite powders, by their mixing and phlegmatization;
- Gelatinous water containing explosives received by a grinding of any types of SRP and powders, are added to a composition as a sensitizer, and consequent next mixing with caliche and thickener. Manufacturing process of receiving of this type of explosive actuates also operation of foster care discharged with special equipment;
- Slab grains of explosive received from missile checkers with diameter of 50-200 mm, used as detonating seismic charges (DSC), originating checkers detonators (CD), monolithic blasthole charges. Technology production process of this type of explosive actuates operations cuttings of charges in definite length, drilling of holes under the pioneers, covering with inert material.

In summary we shall show performances of industrial explosives received from P and SRP (tab. 2).

TABLE 2.

Name of IE	Physical condition	A density G/cm^3	d_g, mm	Velocity of detonation km/sec	TNT equivalent	Packaging
1. Granipores water-resistant (BP-1,BP-3, BM)	Mixture of granulas of artillery double-base and pyroxylin powders by a size 2...20 mm	0,8-0,9	100-120	3,5-3,9	1,09-1,21	Bags
- restrictedly water-resistant (BPS-1, BPS-2, BS)	Mixture granipores BP-1, BP-3 with granulas of ammoniacal caliche	0,85-0,92	110-120	3,8-4,2	0,91-0,93	
2. Non-polluting, water containing (gelpores – 1, 2, 3)	Pasted based fundamentals on pyroxylin powders and SRP and thickened solutions of ammoniacal caliche; granulas up to 3 mm	1,3-1,4	60-90	4,5-5,3	0,85-0,95	bags
3. Charges tubular powder (CTPT)	Oiled beam artillery pyroxylin or mixture pyroxylin and double-base powders		100-130	3,8-4,0	0,84-0,85	Boxes, bags
4. DSC	Cylindrical checkers D=20-120 mm on the basis of double based SRP	1,6-1,65	3-7	6,8-7,5	0,85-0,95	Boxes, bags
5. Charges – detonators	Charges D=70-200 mm on the basis of double based SRP	1,6-1,65	3-5	7,0-7,2	0,85-0,95	Boxes, bags
Charges lide on hollow-charge	Charges D=20-200 mm with cumulative	1,6-1,65	2-10	6,5-7,5	0,85-0,95	Boxes, bags
6. Charges well double base (CDB)	Cylindrical checkers D 200 mm with a point-to point channel. Side and the surfaces are plated by a special material on the basis of double base SRP	1,6-1,65	2-10	7,0-7,5	0,9-1,0	Boxes from a corrugated board bags

Conclusions:
1. Utilization of powders and solid rocket propellants, necessity by which one is called by conversion of defence industry, is the most important state problem aggravated by explosion-hazard and ecological hazard of the product. Solution of this problem lays in a number of problems of liquidation technogenius of catastrophes.

2. The most expedient way of the utilization, allowing major reserves P and SRP, is usage them as industrial explosives. A possibility of regional centres organization including stocks of raw materials (arsenals), of producer and consumer increase an economic efficiency of this way of the utilization.
3. FCDT «Soyuz» has developed efficient ways of operation performance improvement for P and SRP, permitting to use conditions of transport and undermining of industry explosives.
4. On the basis of recovered P and SRP a nomenclature number of industry explosive is developed. Explosives: water-resistant granulouse and gelatinous explosives, detonating seismic charges (DSC), originating checkers - detonators (CD), monolithic blasthole charges (MBC), hollow-charge linear charges (CLC).

APPLICATION OF CRYOCYCLING TO ROCKET MOTOR PROPELLANT SIZE REDUCTION AND REUSE

J. LIPKIN, L.R. WHINNERY, S. GRIFFITHS, R. NILSON
Sandia National Laboratories, Livermore, CA
J. KAMINSKA, G. MOWER, W. MUNSON
Thiokol Propulsion, Brigham City, UT
J. MCNAIR, J. ELLIOTT
General Atomics, San Diego, CA

The Cryocycling technology and its application to processing rocket motor propellant are reviewed in this presentation. Cryocycling is a cost-effective and an environmentally friendly processing method, that can be used to reduce the particle size of many of the solid, bulk propellants obtained from excess munitions. In the Cryocycling process, propellants are repeatedly subjected to cycles of rapid freezing in liquid nitrogen followed by rewarming to near ambient temperatures. After three to four cryocycles, most bulk propellants are reduced to particles in the size range of 5-20 mm without using inherently unsafe cutting operations or generating a secondary waste stream. Further, the energetic properties of the propellant are not altered by Cryocycling, and the resulting particle size range is often appropriate far the direct reapplication and reuse of this processed material as a feed stock for commercial explosives. In the application discussed In this presentation, 360 propellant grains, each weighing 10.9 kg, are batch processed each day in an automated Cryocycling processing facility located at Thiokol Propulsion, Brigham City. UT. These grains are roughly 1 m long and are made of a double t3ased propellant designated JPIV; they hr3ve a cruciform cross section roughly 0.12 m in diameter. The cryocycled product is packaged and shipped to a commercial explosives producer for reapplication in their products.
This work was supported by Sandia National Laboratories. Livermore, CA. Sandia is a multiprogram laboratory operated by Sandier Corporation. a Lockheed Martin Company, for the United States Department of energy under Contract DE-AC04-94AL8500.

COFIRING OF PROPELLANT WASHOUT RESIDUE WITH TRADITIONAL BOILER FUELS: RESOLUTION OF OPERATIONAL AND ENVIRONMENTAL ISSUES

S. G. BUCKLEY[†], J. LIPKIN, L. L. BAXTER, R. MOEHRLE,
J. R. ROSS, G. MOWER[‡], AND W. MUNSON[‡],
[†]*University of Maryland, College Park, MD 20742*
Sandia National Laboratories, Livermore, CA 94551
[‡]*Thiokol Propulsion Division, Brigham City, UT 84302*

Abstract

High-pressure water washout of large solid rocket motors yields a residue, which is primarily comprised of aluminum flake, polybutadiene rubber, ammonium perchlorate, and asbestos entrained from the liner. Treatment following washout recovers most of the ammonium perchlorate from the residue for reuse, desensitizing the washout residue. Often this material is subsequently disposed of through open burning/open detonation (OB/OD) or in landfills.

Cofiring combustion is a compelling alternative disposal option for desensitized materials of this type. In the cofiring process, the washout residue is mixed with conventional boiler fuels, and standard boiler pollution control equipment is used to mitigate environmental hazards. This paper summarizes the findings of research into both the operational and environmental considerations surrounding cofiring combustion of washout residue. This material was fired in two different combustors: in a down-fired research combustor with electrically heated walls and a small-scale bed combustor. Measurements of asbestos, chlorinated species, and temperature were made to determine combustion characteristics. Under a wide range of conditions that are easily obtained in combustion systems, it was found that: (1) asbestos in the feed is rendered non-hazardous, (2) gas-phase combustion products show no evidence of toxic chlorinated combustion products or precursors, and (3) chlorine removal from the ash scales with the extent of fuel burnout. In addition, measured bed temperatures during combustion of the highly aluminized fuel never exceeded 450 °C, while measured bulk temperatures in the fuel pile remained below 900 °C. Hence, despite the fact the aluminum in the fuel has an adiabatic flame temperature in excess of 2500 °C, there is little potential for damage to the combustion system.

1. Introduction and Background

Millions of pounds of waste energetic materials (EM) such as rocket motor propellant and explosives are generated every year. While some of the material can be recycled, much of it is difficult to separate or has little intrinsic value, and some of the material contains additional hazards such as asbestos and chlorine. Typically this material is desensitized and then disposed of through incineration, open burning/open detonation (OB/OD) or, if nonreactive, in landfills. None of these disposal methods takes advantage of the energy content of the fuels and they may generate significant pollution. Combustion in an industrial furnace, boiler, or kiln (hereafter referred to jointly as "industrial furnaces") is a compelling alternative disposal option for desensitized EM. This process can be used to recover energy value from the fuel and also can serve to mitigate many environmental hazards associated with other options through the use of furnace pollution controls.

At present the majority of EM disposal is by landfill interment or OB/OD. Landfilling of desensitized EM leaves an unwanted legacy for future generations, with the concurrent risk that chlorine and other constituents may leach into the surrounding soil and contaminate groundwater. OB/OD is typically accomplished at military installations removed from the public. Despite the relative isolation of these sites, detonation shock waves have been known to reflect off of the upper atmosphere and break windows in towns 10-20 miles distant [1], and plumes from the detonation and burning are typically visible for many miles. Recent measurements of emissions from OB/OD of antipersonnel mines indicate that significant quantities of pollutants can be emitted; measured emission factors (mass of pollutant/mass of EM) were 5.84×10^{-3}, 8.15×10^{-4}, and 3.06×10^{-3} for methane, benzene, and total aromatics, respectively [2]. In measurements from contained burns of solid rocket motors, the exhaust contained roughly 3% HCl, 100 ppm HCN, 2,500 ppm CH_4, and 100 ppm of volatile organics (such as chloromethane, benzene, and 1,2 dichloroethane) on a mass basis [3]. In addition, NO_X emissions from uncontrolled combustion of EM can be quite high due to the high nitrogen content of the EM. Because of public and regulatory concerns, both landfilling and OB/OD of these materials are becoming increasingly unacceptable.

Given the combination of environmental concerns and disposal costs of approximately $850 per metric ton [4], reapplication of EM is highly desirable. A preliminary assessment of the combustion properties of desensitized energetic materials has shown that the concept holds promise [5], but no pilot-scale investigation has been completed. Cofiring the energetic materials with standard fuels affords the advantage of pre-existing pollution control equipment to mitigate any environmental hazards, as well as the benefit of energy recovery from the energetic material-derived fuels (EMDF). In the cofiring scenario(s) that we envision, EM are desensitized and transformed into EMDF, that would be fired in a 1-5% blend ratio with fossil fuels in an furnace.

The subject of the investigation is residue from high-pressure water washout of solid rocket motors, in which high-pressure water jets are used to remove propellant material from the casing of a rocket. The high-pressure water also removes some of the insulation of the liner, which often contains asbestos. Treatment following washout recovers most of the ammonium perchlorate from the residue for reuse. The nominal composition of the material before and after the washout process is shown in Table 1. In this paper the material following washout is referred to either as washout residue or energetic material-derived fuel (EMDF). Because of the asbestos and residual ammonium perchlorate content, this material must be buried in a permitted landfill or burned under controlled conditions. These two components also introduce issues for a combustion disposal option, specifically the fate of the asbestos and the potential production of unwanted chlorinated byproducts. In addition, the presence of aluminum powder in the residue results in a measured combustion temperature of over 2000 °C for particles of approximately 1 mm in size. Concern over potential damage to furnace equipment from the high temperature material is therefore an issue. Finally, the effect of the residue on the ash stream from the furnace must be determined, if the ash has value in a particular application (e.g. cement additive, asphalt, fertilizer, etc.).

When this study of reapplication options for washout residue was initiated, the primary concern was the presence of 0.2 – 1% asbestos in the residue. Asbestos is defined as any one of six naturally-occurring silicate compounds (chrysotile, riebeckite, and grunerite are common). Asbestos is considered a human health hazard when it has a fibrous shape that is more than 5 microns in length, less than 3 microns in diameter and has a 3:1 or greater aspect ratio [6]. The inhalation danger from asbestos fibers results from the fact that the fibers are insoluble, hard, and nearly impossible for the body to remove from the lung. The fibrous shape causes irritation to the lung that may produce lung cancer, asbestosis, or mesothelioma, a rare type of cancer associated with asbestos exposure [7].

All of the forms of asbestos melt at or below 1400 °C: grunerite melts at 1399 °C, riebeckite melts at 1193 °C, and chrysotile begins to soften at 600 °C, and is fully changed in chemical form below 1000 °C. Chrysotile is the most common form of asbestos because it is easily spun; it accounts for roughly 95% of the combined U.S. and Canadian asbestos production, and is the type of insulation used in strategic rocket motors. The melting temperature of all of these asbestos forms, particularly chrysotile, is low compared with the combustion temperature of the aluminum-containing fuel. Our initial hypothesis was that combustion of the residue would melt the asbestos fibers. Melting the fibers changes their shape, rendering them non-hazardous.

This paper discusses our findings concerning asbestos mitigation, the formation of chlorinated air toxics, the removal of chlorine from the ash, and high-temperature material-related issues during combustion experiments with washout material in Sandia's MultiFuel Combustor (MFC) and in a small bed combustor. This data builds upon previously published work in this area [8].

TABLE 1. Nominal composition of solid fuel and rocket motor washout residue.

Component	Approximate composition (percent)	
	Solid fuel	Washout residue (dry basis)
Ammonium perchlorate	65%	5%
Aluminum powder	20%	55%
Polybutadiene rubber binder	15%	40%
Asbestos (entrained from liner)	n/a	up to 1%

2. Experimental Methods

The MFC is illustrated in Fig. 1. The MFC is a pilot-scale (30-100 kW, depending on fuel type) facility that simulates the local environment to which independently injected solid, liquid, or gaseous fuels are exposed as they pass through an entrained-flow combustion system. Fuels are introduced directly into the silicon-carbide-lined combustion section, where they may form a self-supporting flame. The MFC has an integral preheat burner (natural gas/air) which was used in some of these experiments. Combustion products are extracted from the combustor with a heated line to prevent water condensation, filtered by a heated filter, and quantified using NDIR NO_X, CO, CO_2, and SO_2 analyzers and a paramagnetic O_2 analyzer, all from Horiba Instruments (Irvine, CA). Analyzers are calibrated through the entire sampling system prior to each experiment.

A separate, identical heated sampling line transfers products to a heated 24-meter long-path gas cell, which is mated to a Mattson Instruments (Madison, WI) Infinity Fourier Transform Infrared (FTIR) spectrometer with 0.5 cm^{-1} resolution. Most spectra in these experiments were taken with the cell at 110 °C and a cell pressure of 300 torr. The reduced pressure was an extra precaution to prevent condensation of volatile hydrocarbons and water in the sample lines and in the cell, particularly in case areas of the cell and flow system were not uniformly heated. Sample continuously flowed through the cell during measurements to reduce the chance of adsorption and/or condensation of species in the cell and sample lines perturbing the concentrations of species in the sample. The flow in the transfer lines and cell was allowed to come to steady state for five minutes prior to recording a spectrum. Between samples, the cell was completely evacuated at least twice, and the cell and sample lines were flushed with dry nitrogen for at least 10 minutes. Under these conditions, detection limits of most hydrocarbons in the cell should be roughly 10 parts-per-million (ppm) on a molar basis, based on earlier work with butadiene and methane. As there was no need to quantify our findings, calibrations were not performed for these experiments. Additional description of the MultiFuel Combustor facility can be found in the literature [9].

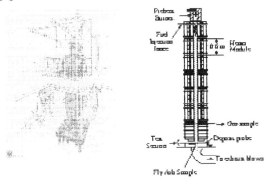

Figure 1. Artist's rendition (left) and schematic diagram (right) of the MultiFuel Combustor.

The data reported here were obtained by injecting test EMDF in the top section of the MFC and sampling combustion products from the top of the 7th section. The gas residence time was 1.3 – 2 seconds, depending on the combustor wall temperatures, which were varied from 700 – 1200 °C. The EMDF was fed using a custom screw-auger feeder developed by Thiokol, which was loaned to Sandia. The EMDF fuel itself (hereafter occasionally referred to simply as "fuel") is granular, with a characteristic size between 3 and 8 mm. It had been treated by Thiokol to remove the moisture and the tendency of the particles to re-agglomerate.

For the experiments reported here, fly ash sample was collected using a 333 micron mesh screen in a 4-inch diameter probe inserted into the flow downstream of the combustor where the temperature was approximately 300 °C. The collection device intercepted roughly 1/3 of the combustor flow, and would remain in the flow for 10 minutes, which was long enough to capture sufficient sample for analysis. During these experiments the gas flow in the combustor was 20 scfm, and the solid fuel feed was approximately 25 g/min. For a 900 °C gas temperature this yields an approximately 1.8-second gas residence time.

Figure 2 shows a schematic of the bed combustor assembled for these experiments. House air was injected at a velocity corresponding to 1-2 ft/s at the bed. Premixed methane / air burners fired over the bed, with the flame sweeping down on to the bed material, causing ignition. Fuel placed on the bed was 10, 20, and 50% energetic material by energy value, blended with biomass (pine) or coal. Samples of the fuel and post-combustion ash were analyzed for chlorine content, and temperatures throughout the bed, fuel, and exhaust system were measured for the duration of the runs (approximately 20 minutes) using type R thermocouples.

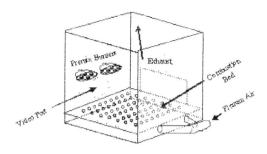

Figure 2. Schematic bed combustor

3. Results

3.1. POTENTIAL FOR FORMATION OF AIR TOXICS

FTIR analysis was performed on gas samples of combustion products generated under a variety of temperature and flow conditions, ranging from 700 – 1200 °C and from fuel rich to 12% excess O_2. These conditions represent a spectrum of operating conditions that might be encountered in a commercial furnace that could be used to cofire the solid rocket motor washout material. Typically, industrial furnaces are fired with at least 5% excess oxygen and have a background (radiative) temperature in the range 700 – 1200 °C. Gas temperatures are usually much higher; for example, exit temperatures at a cement kiln under consideration are reported to be 1650 °C. The combination of high temperatures, sufficient oxygen, and sufficient residence time assures good combustion with a minimum of unwanted toxic byproducts.

In the combustion systems considered for cofiring, bulk temperature and residence time are well characterized and controlled. Typically, emissions of hazardous byproducts from these devices are the result of poor mixing between the fuel and the air, which can result in a hot, fuel-rich pocket of gas escaping from the combustion section. To determine the possibility of air toxic emissions from such incomplete combustion events, these tests also included extreme cases of low-oxygen (less than 1%). These tests are notable for the measurable CO in the product gases, which is a common indicator of incomplete and inefficient combustion.

Figure 3 shows the FTIR spectrum from the combustion of rocket motor washout material with wall temperatures of 700 °C and 10% excess O_2. This spectrum is typical of almost all of the spectra taken in the test series. The H_2O and CO_2 bands are saturated, and features due to HCl and CO are the only resolvable features in the spectrum. Figures 4 and 5 show the 700 – 1200 cm^{-1} and 1900 – 3200 cm^{-1} spectral regions, respectively. No features were found in any of the spectra that suggest the presence of chlorinated air toxics. The detection limit of the system for these species is 5 – 10 ppm under these conditions. The spectra were not only checked with the on-line Mattson libraries mentioned above, but were also compared with printed versions of high-resolution spectra of over 50 hazardous chlorinated combustion byproducts from the chlorinated hydrocarbon combustion research laboratory at U.C. Berkeley [10]. The results indicate that the only gas-phase chlorinated species detected under all the conditions tested is HCl, which can be easily scrubbed from the exhaust stream.

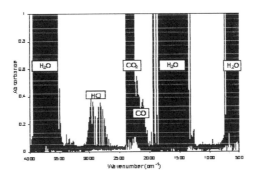

Figure 3. Typical FTIR absorption spectrum from combustion of rocket motor washout residue. This spectrum corresponds with wall temperatures of 700°C and 10% excess O_2.

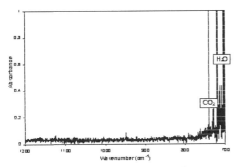 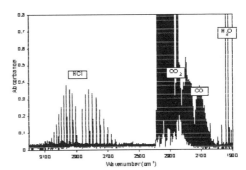

Figure 4. Enlarged fingerprint region from Figure 3.

Figure 5. Enlarged region 1900 – 3200 cm^{-1} from Figure 3

Only in the coolest (700 °C walls), fuel-rich condition did we observe any gas-phase combustion products besides H_2O, CO_2, HCl, and CO. Figure 6 shows the interesting features in the spectrum corresponding to this experiment. Besides HCl, we observe evidence of CH_4, an aromatic type =C-H stretch bond, and OH/NH stretch bonds. These features are derived from fuel components. In the fuel-rich condition, methane could either be formed from incomplete hydrocarbon breakdown products, such as CH_2 and CH_3 radicals, or could come directly from unreacted natural gas from the preheat burner. Below the methane spectra, which consists of a tall peak (Q branch) surrounded by narrow features on either side (P and R branches), there is a broad absorption feature from 3000 – 3100 cm^{-1}. This feature is due to a double-bonded carbon atom bonded to a hydrogen atom: =C-H, commonly seen associated with aromatic molecules such as benzene and toluene. In this case however, it is most likely due to the binder material. Polybutadiene is a linear chain molecule with a structure of -HC=HC-HC=HC-. This structure is identical to that of a broken aromatic ring, and hence the spectral feature is shared. Hence this broad feature is probably due to incomplete breakdown of the polybutadiene in the fuel- rich combustor.

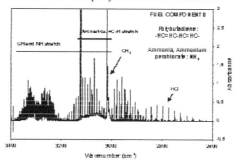

Figure 6. Underoxidized fuel components observed at 700 °C, fuel-rich conditions.

Finally, the remaining spectral feature between 3200 – 3400 cm^{-1} is in a region associated with NH and OH stretching vibrations. As part of the fuel preparation step, Thiokol treated the fuel with ammonia to eliminate the tendency for re-agglomeration. The spectral feature between 3200 – 3400 cm^{-1} is almost certainly from an amine (NH) bond formed in the postcombustion zone due to incomplete reaction of either the ammonium perchlorate (NH_4ClO_4) or the ammonia in the fuel.

3.2. ASBESTOS REMEDIATION

Fly ash was collected from the MFC during EMDF combustion experiments, over a wide range of experimental conditions. The majority of the samples were collected using a 4-inch diameter probe into which a 333-micron mesh screen was inserted. The funnel was placed in the combustion flow for 10 minutes, which was long enough for the fly ash to cover the screen completely. Following data collection, some samples were sent to Huffman Laboratories (Golden, CO) for proximate, ultimate, and ash analysis. All samples were sent to Thiokol's analytical lab (Brigham City, UT) for asbestos analysis in their OSHA-certified laboratory.

Table 2 shows the results of the asbestos analysis of the ash from the MFC. Asbestos fibers were found in the products only when the wall temperatures were 900 °C or lower. Significantly, even in these cases, asbestos was never found free-floating in the collected deposit, but was always found embedded in larger, rubber particles. Hence even when present, the asbestos from the residue would not be an inhalation hazard.

TABLE 2: Experimental results showing asbestos conversion (to aluminum/magnesium silicates) at combustor temperatures of 1000 °C and above.

Experimental result	REACTOR WALL TEMPERATURE						
	600°C	700°C	800°C	900°C	1000°C	1100°C	1200°C
Asbestos found in unburned rubber particles in ash	YES	YES	YES	YES	NO	NO	NO
Free-floating asbestos in ash	NO	NO	NO	NO	NO	NO	NO
Heat-conversion of asbestos occured	NO	NO	NO	NO	YES	YES	YES
Carbon content of ash (percent)	8.6	1.55	1.44	0.85	1.42	0.49	

At wall temperatures of 1000 °C and above, the asbestos transforms into harmless, amorphous particles that appear relatively spherical; no asbestos fibers were found in the collected deposit.

3.3. TEMPERATURES IN BED COMBUSTOR

Of primary concern in the practical application of cofiring of washout residue is the maximum temperature achieved in a combustor burning this heavily aluminized fuel. As noted above, the adiabatic flame temperature of aluminum is over 2500 °C, which is far above the damage threshold of any material used in construction of industrial boilers. Experiments in the bed combustor were designed to determine the allowable mixture ratio

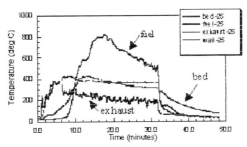

of EMDF to boiler fuel that could be used in a system with a stainless steel bed, and a mild steel bed.

Figure 7: Temperature / time profiles for 75% biomass / 25% EM blend.

Surprisingly, maximum temperatures in the bed, bulk fuel, and exhaust all remained well within limits of current materials for all mixture ratios. Figure 7 illustrates the temperature / time history for a mixture of 25% EMDF / 75% pine (by energy content). As shown in the figure, the maximum temperature experienced by the bed is approximately 400 °C, as measured in the center of the bed under the fuel pile using a type R thermocouple. Apparently conduction through the bed is rapid enough that any hot spots are quickly mitigated. No material damage was observed either on the steel bed, or on the 316 stainless steel bed. Similarly, the maximum temperatures in the fuel pile and in the exhaust were well controlled, and very close to those found in typical bed combustion systems.

3.4. HALOGEN PERSISTENCE IN ASH

The final obstacle identified to adoption of cofiring EMDF relates to the residual chlorine content of the ash following combustion. Much combustion ash is used in concrete mix and other industrial applications. High chlorine content makes the ash unsuitable for many of these applications, and thus a high degree of halogen persistence in the ash fraction would make cofiring combustion unattractive for industrial facilities.

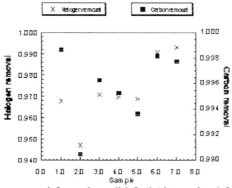

Figure 8: Halogen removal from the solid fuel (determined from ash analysis) correlates well with carbon removal.

Tests for total halogen were completed on the mixed fuel samples and on the residual ash by Huffman Laboratories. Figure 8 illustrates the results. Halogen removal is shown to correlate well with carbon removal. As most combustion systems are designed to achieve optimal energy release, they are optimized for good fuel burnout; hence residual chlorine in the ash should not be a problem.

4. Conclusions

Based on the findings of this investigation, we conclude that the combustion properties of rocket motor washout material are positive in terms of the potential for cofiring as a recycle

/ reuse option to landfill disposal. Using experimental data derived from Sandia's MFC and a bed combustor; we have shown that the major potential issues associated with combustion of EMDF as a cofiring fuel can be mitigated provided good combustion practice is followed.

First, fuel analysis shows that between 2 and 4 percent by mass of the residue is chlorine, which could pose a hazard due to the formation of chlorinated air toxics. However, no chlorinated air toxics were found at any combustion condition (fuel-rich to 10% excess O_2, 700 – 1200 °C combustor walls). All of the chlorine appears to be converted to HCl, which can be readily scrubbed from the exhaust gases.

Second, no respirable asbestos fibers were found in the fly ash under any condition tested (wall temperatures 500 – 1200 °C, natural gas preheat combustor on and off). Asbestos fibers were found bound in unburned rubber pieces when the combustor wall temperatures were 900 °C and below, particularly when the preheat combustor was on. We note that peak temperatures in all of the industrial combustion devices under consideration exceed 900 °C by a comfortable margin.

Third, peak combustion bed temperatures in all cases were below 900 °C, and bed temperatures were below 500 °C. No damage was observed on the replaceable beds that were used in the experimental bed combustor.

Finally, halogen residual in the ash was shown to correlate well with residual carbon. As energy systems are typically optimized for carbon conversion into energy, this suggests that halogen remaining in the ash should not be a large problem.

These observations suggest that good combustion practice in a commercial furnace or kiln facility is likely to guarantee mitigation of both the air toxic and asbestos hazards that are potentially associated with cofiring solid rocket motor washout residue. We also believe that materials issues in the combustor should not pose a problem. Furthermore, it is likely that residual chlorine in the ash will prove useful in recycling applications such as concrete additives, if carbon burnout in the bulk fuel is complete. For these reasons, we believe cofiring rocket motor washout residue with conventional boiler fuels is a viable alternative to other disposal options for this material.

Acknowledgements
The authors gratefully acknowledge funding from the joint Department of Defense, Department of Energy Memorandum of Understanding addressing conventional munition technology development. Special encouragement provided by Mr. James Q. Wheeler, Director, US Army Defense Ammunition Center, is also acknowledged.

References

1. Personal Communications with Sierra Army Depot Staff (1996).
2. Wilcox, J.L., (1996) 4th Global Demilitarization Symposium, pp. 388-399.
3. Steele, D. (1996) (1996) 4th Global Demilitarization Symposium, pp.135-149
4. Arbuckle, J., (1996) 4th Global Demilitarization Symposium, pp.13-22
5. Myler, C.A., Bradshaw, W.M. and Cosmos, M.G., (1991) *Journal of Hazardous Materials* **26**, 3333-342.
6. Lewis, J.R. Sr., (1994) *Rapid Guide to Hazardous Chemicals in the Workplace, 3rd Ed.* Van Nostrand Reinhold, New York, p.17.
7. Amdur, M.O., Doull, J. and Klassen, C.D., (1991) *Casarett and Doull's Toxicology: The Basic Science of Poisons, 4th Ed.*, McGraw-Hill, New York.
8. Buckley, S.G., Robinson A.L., Baxter, L.L., (1998) *Energetics to Energy: Combustion and Environmental Implications Surrounding the Reapplication of Energetic Materials as Boiler Fuels*, 27th International Symposium on Combustion, Boulder, CO.
9. Baxter, L.L., (1992) *Combustion and Flame* **90**, 174 – 184.
10. Courtesy Professors Robert .F. Sawyer and Catherine P. Koshland and graduate student Lee Anne Sgro.

CONVERSION OF DEMILITARIZED EXPLOSIVES AND PROPELLANTS TO HIGHER VALUE PRODUCTS

A.R. MITCHELL, M.D. COBURN, R.D. SCHMIDT, P.F. PAGORIA
AND G.S. LEE
Lawrence Livermore National Laboratory
Energetic Materials Center, MS L-282, P.O. Box 808
Livermore, CA 94550

1. Introduction

The objective of this project is to develop new and innovative solutions for the disposal of surplus energetic materials. Disposal through open burning/open detonation (OB/OD) is less attractive today due to environmental, cost and safety concerns. We are examining the use of military high explosives as raw materials for the production of higher value products useful in civilian and military applications. We have developed scenarios where Explosive D and TNT can be used as raw materials for industrial processes to produce higher value products [1,2]. We describe in this report our progress in obtaining 1,3,5-triamino-2,4,6-trinitro-benzene (TATB), a higher value explosive, using starting materials from demilitarization inventories (Explosive D) or low-cost commodity chemicals (4-nitroaniline).

TATB is a reasonably powerful high explosive (HE) whose thermal and shock stability is considerably greater than that of any other known material of comparable energy [3]. The high stability of TATB favors its use in military [4] and civilian applications [5] when insensitive high explosives are required. In addition to its applications as a HE, TATB is used to produce the important intermediate benzenehexamine [6-10]. Benzenehexamine has been used in the preparation of ferromagnetic organic salts [10] and in the synthesis of new heteropolycyclic molecules such as 1,4,5,8,9,12-hexaazatriphenylene (HAT) that serve as strong electron acceptor ligands for low-valence transition metals [7,9]. The use of TATB to prepare components of lyotropic liquid-crystal phases for use in display devices is the subject of a German patent [11].

There is a definite need for a less expensive and more environmentally benign production of TATB. Current production techniques for making TATB are expensive and rely on environmentally hazardous reagents and relatively harsh conditions. We recently reported a novel approach to the synthesis of TATB which utilizes relatively inexpensive starting materials and mild reaction conditions [2,12,13]. This new process relies on amination of nitroaromatic starting materials using a reaction known as Vicarious Nucleophilic Substitution (VNS) of hydrogen [14]. Figure 1 outlines the approach. We

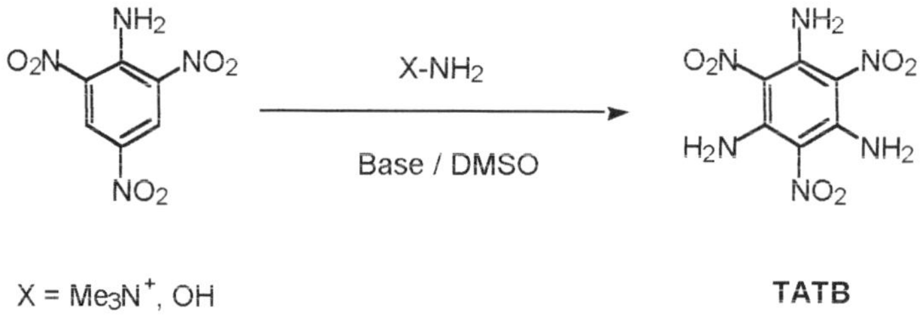

X = Me$_3$N$^+$, OH   TATB

Figure 1. VNS synthesis of TATB from picramide.

have been working on the scale-up of this new synthesis with the goal of developing a new production of TATB.

2. Process Studies with 1,1,1-Trimethylhydrazinium Iodide as the VNS Aminating Reagent

2.1. INITIAL STUDIES

We have determined that 1,1,1-trimethylhydrazinium iodide (TMHI) is the most efficient aminating reagent available for the VNS synthesis of TATB [2,12,13,15]. Picramide and solid TMHI are dissolved in DMSO, and base (sodium methoxide or ethoxide) is added to initiate the reaction. The reaction is conducted at room temperature, and is complete in under 3 hours, giving TATB in 80-90% yield (Figure 2). The major expected impurity is 1,3-diamino-2,4,6-trinitrobenzene (DATB), which results from incomplete amination. Under these reaction conditions, no DATB (≤0.5%) was detected by FTIR spectroscopy or direct insertion solids probe mass spectrometry (DIP-MS).

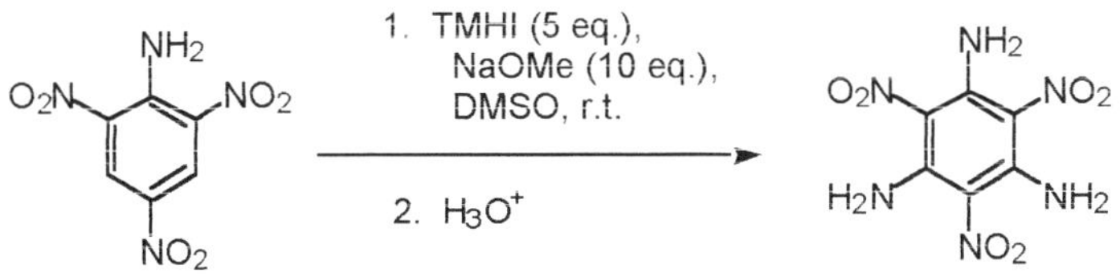

Figure 2. VNS synthesis of TATB using TMHI

2.2. STUDIES ON VARYING REACTION CONDITIONS

The initial studies of this reaction employed picramide concentrations ≤ 0.13 M with large excesses of TMHI and base to drive the reaction to completion. We examined the effects of decreased solvent and reagents on the reaction. In general, the reaction will run efficiently up to 0.2 M picramide and using 3 eq. TMHI. The success of the reaction seems most dependent on base, requiring 8 eq. to proceed efficiently. The yield and purity of TATB drop significantly if an insufficient excess of base is used. It was also found that the reaction is very sensitive to the quality of the base, particularly in the case of sodium methoxide: older lots of the base which had been exposed to air, even while retaining the identical physical appearance of fresh material, gave reduced yields (or, in the worst case, no yield at all) of TATB. Thus far, the largest scale attempted has been the 10 gram level. The reaction appears to scale linearly, delivering 82% yield of TATB at >99% purity. Larger scale work is currently in progress.

2.3. METHODS OF QUENCHING THE REACTION

All initial studies of this reaction used either aqueous mineral acid solutions or water to quench the reaction and induce precipitation of TATB. This method results in a very small particle size, on the order of 0.2-1 μm. It was reasoned that quenching with a weak organic acid in the absence of water might result in larger particle size. We found that quenching with citric acid monohydrate in DMSO produced particles in the 1-10 μm range. A larger particle size (30 μm) has recently been obtained using other organic acids [16]. It was also noticed that the final color of the product TATB varied when different quenching solutions were used.

2.4. IN SITU GENERATION OF TMHI

Although TMHI is easy to prepare and handle [13,15] its use in the solid form requires an additional synthesis and isolation step, which would increase the overall product costs at production scale. Therefore, several experiments were conducted which examined the *in situ* generation of the reagent. To accomplish this, the precursor reagents--1,1-dimethylhydrazine and methyl iodide--were sequentially added to DMSO and allowed to react. Picramide was then added to this solution, followed by base, and the reaction was allowed to proceed as before. This method appears to give at least as good results as the original method, and in several cases gave slightly higher yields of TATB.

2.5. QUALITY OF STARTING MATERIALS

As mentioned earlier, the reaction appears to be very sensitive to the condition of the sodium methoxide. Several attempts at making TATB using an older lot of NaOMe failed, even though the base had been stored in a dessicator and the physical appearance of the base was no different from newer material (white, free-flowing fine powder). Analysis of this lot of NaOMe revealed that much of it had been converted to sodium carbonate by absorption of ambient CO_2 which inactivated it in the VNS reaction. Good yields (>85%) were again obtained when fresh NaOMe was employed.

In a few experiments, there was some variation in the purity of the starting picramide, and this appears to have affected the final appearance of the TATB, even though the TATB appears to be chemically >99% pure by spectroscopy. The principal impurity in the picramide was picryl chloride (vide infra). Impurity levels of greater than a few percent cause the product TATB to darken and, of course, reduce the total yield of TATB (although corrected yields are similar to those using pure picramide). High levels of impurities in starting picramide also change the crystal morphology of the product TATB.

2.6. PRODUCT ANALYSIS

Since TATB is nearly insoluble in most solvents, simpler forms of chemical analysis such as NMR or Gas Chromatography are not practicable. Therefore, other techniques which allow analysis of the solid were investigated. The first of these employed was Fourier Transform Infrared Spectroscopy (FTIR). The amine N-H stretching modes in TATB produce two characteristic absorptions at approximately 3225 and 3325 cm^{-1}, while those for DATB occur at 3360 and 3390 cm^{-1}. By using Nujol mull preparations for TATB samples, we have found that DATB can be reliably detected at concentrations of 1% or greater.

Another technique for TATB product analysis which we are using is direct insertion solids probe mass spectrometry (DIP-MS). In this technique, a solid sample of TATB is placed in a sample holder at the end of a probe. The probe tip is inserted into a mass spectrometer, and is heated to cause the solid sample to evaporate into the MS ion volume, thereby allowing analysis of solids. Compounds with differing volatilities will evaporate at different times (a process known "probe distillation") and can thus be resolved to some extent by the MS detector. We have found that DATB can be reliably detected in a TATB sample at 1% concentration, and in some cases in concentrations as low as 0.1%.

Selected samples were submitted for elemental analysis. In early samples, the elemental analysis revealed that the product TATB was contaminated with 1-2% iodine. Unreacted TMHI as a source of iodine contamination was ruled out as TMHI could not be detected in the TATB samples using mass spectroscopy. We are examining the effects of quenching methods on impurities such as iodine, chlorine, etc.

In order to compare the TATB from this VNS process to that from more traditional processes, we have also conducted DSC, CRT, DH_{50}, spark and friction sensitivity tests on this material. In general, results are similar to those observed for TATB, except that thus far, the DSC values run consistently low by about 20-30 degrees. This may be an artifact of the much finer particle size produced by this process, although to confirm this more tests will be needed.

3. Process Studies with Hydroxylamine as the VNS Aminating Reagent

Due to the relative toxicity and cost of reagents used to make TMHI, we reinvestigated the use of hydroxylamine as a VNS aminating reagent. Hydroxylamine is in fact, the earliest known example of a VNS aminating reagent [17] although the term "VNS" was coined many decades later [14]. Our earliest work in aminating picramide with hydroxylamine was disappointing since the reaction only provided DATB containing trace amounts of TATB at best [12]. The poor reactivity of hydroxylamine was independently confirmed by Seko and Kawamura who were unable to aminate nitrobenzene using hydroxylamine [18].

The low cost of hydroxylamine as an aminating reagent warranted further investigation and recent work in our laboratories showed that hydroxylamine will aminate picramide to TATB at elevated temperature (65-90°C) to furnish TATB (Scheme 3) [19]. Although the

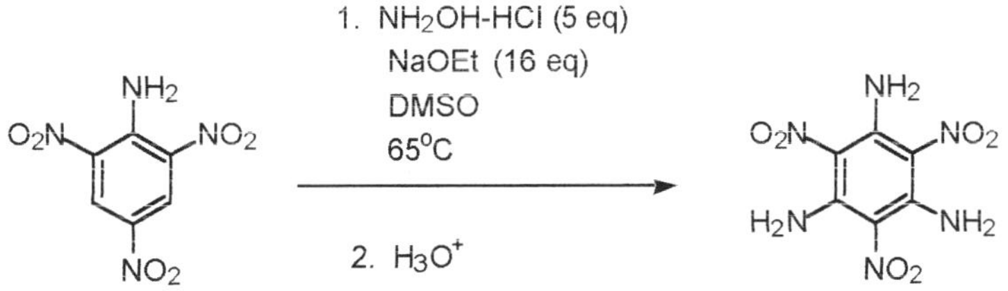

Figure 3. VNS synthesis of TATB using hydroxylamine hydrochloride

work with hydroxylamine is preliminary, satisfactory yields of TATB at approximately 97% purity have been achieved. Thus far the best results were obtained using NaOEt as the base in DMSO at 65°C for 6-12 hours. We are in the process of testing other hydroxylamine salts and anticipate the purity of the product will increase to over 99%. The relatively low cost of hydroxylamine salts makes this option very attractive.

4. Studies of Picramide Synthesis

4.1. NITRATION OF 4-NITROANILINE

Picramide is no longer commercially available. Therefore, as part of this project, we were required to reinvestigate methods for its production. One simple method is nitration of 4-nitroaniline, an inexpensive commodity chemical (Figure 4) [20]. Early studies in

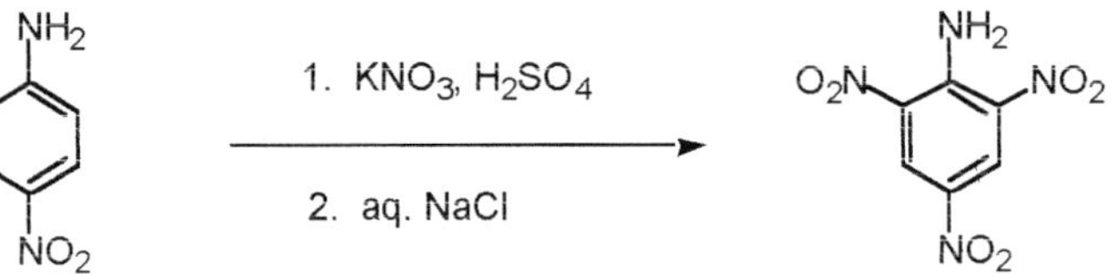

Figure 4. Synthesis of picramide

our laboratories using similar conditions gave good results, although some impurities were noted, the most significant being picryl chloride. (The workup of picramide is facilitated by the addition of brine, which apparently gives rise to the picryl chloride impurity.) In one case, picryl chloride was present in up to 20% impurity. Such impurities would require expensive recrystallization processing, since they affect the quality of TATB produced, as discussed earlier. However, our project collaborators at Pantex (Mason & Hanger Corporation, Amarillo, Texas) have improved the process and have prepared picramide in high yields (90%) and purity (>99.5%).

4.2. CONVERSION OF EXPLOSIVE D TO PICRAMIDE

Although the direct conversion of picric acid to picramide is known [21], an analogous conversion of Explosive D to picramide has not been reported. We are evaluating many routes to picramide from Explosive D, including the conversions illustrated in Figure 5, with respect to cost, convenience and safety [16].

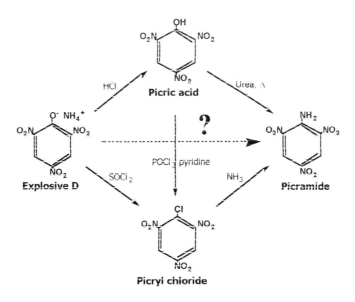

Figure 5. Conversion of Explosive D to picramide

5. Conclusions

Chemical conversions of surplus energetic materials (high explosives, propellants) offer constructive alternatives to destruction. Explosives containing nitroarenes (TNT, ammonium picrate) are especially useful as feedstocks for industrial processes that produce higher value products.

6. Acknowledgements

Work performed under the auspices of the U.S. Department of Energy by Lawrence Livermore National Laboratory under Contract No. W-7405-ENG-48. We wish to acknowledge contributions to this project by Messrs. Ray Thorpe, W. Tim Quinlin, Patrick Phelan and Monty Cates (Mason & Hangar Co., Pantex Plant, Amarillo, Texas, USA).

Funding for this work has been provided by James Q. Wheeler (Director, U.S. Army Defense Ammunition Center) and the Joint DOD/DOE Munitions Technology Program, Munitions Life Cycle Technology R&D Program (Technology Coordination Group IX) with matching funds from DOE (ADAPT Program).

7. References

1. Mitchell, A. R. and Sanner, R. D. (1993) Chemical conversion of energetic materials to higher value products, in H. Ebeling (ed.) *Energetic Materials- Insensitivity and Environmental Awareness*, Ebeling, Proceedings of the 24th International Annual Conference of ICT, Karlsruhe, Germany, pp. 38: 1-6.
2. Mitchell, Alexander R.; Pagoria, Philip F.and Schmidt, Robert D. (1996) A new synthesis of TATB using inexpensive starting materials and mild reaction conditions, in T Keicher (ed.) *Energetic Materials-Technology, Manufacturing and Processing*, Keicher, T., Ed., Proceedings of the 27th International Annual Conference of ICT, Karlsruhe, Germany, pp. 29: 1-11.
3. Rice, S. F. and Simpson, R. L. (1990) The unusual stability of TATB: a review of the scientific literature, Lawrence Livermore National Laboratory, Livermore, CA, Report UCRL-LR-103683.
4. Dobratz, B.M. (1995) The insensitive high explosive triaminotrinitrobenzene (TATB): development and characterization - 1888 to 1994, Los Alamos Scientific Laboratory, Los Alamos, NM, Report LA-13014-H.
5. Voreck, W.E., Brooks, J.E., Eberhardt, J.R. and Rezaie, H.A. (1997) Shaped charge for a perforating gun having a main body of explosive including TATB and a sensitive primer, *U.S. Patent 5,597,974.*
6. Kohne, B. and Praefcke, K. (1987) Isolierung farblosen Benzolhexamins, *Liebigs Ann. Chem.*, 265.
7. Rogers, D.Z. (1986) Improved synthesis of 1,4,5,8,9,12-hexaazatriphenylene, *J. Org. Chem.*, **51**, 3904-3905.
8. Kohne, B., Praefcke, K., Derz, T. Gondro, T. and Frolow, F. (1986) Benzotri-(imidazole) - a new ring system derived from benzenehexamine, *Angew. Chem. Int. Ed. Engl.*, **25**, 650-651.
9. Nasielski-Hinkens, R., Benedek-Vamos, M., Maetens, D. and Nasielski, J. (1981) A new heterocyclic ligand for transition metals: 1,4,5,8,9,12-hexaazatriphenylene and its chromium carbonyl complexes, *J. Organomet. Chem.*, **217**, 179-182.
10. Breslow, R., Maslak, P. and Thomaides, J.S. (1984) Synthesis of the hexaamino-benzene derivative hexaazaoctadecahydrocoronene (HOC) and related cations, *J.Am. Chem. Soc.*, **106**, 6453-6454.
11. Praefcke, K. and Kohne, B. (1988) Amido compounds as components of lyotropic liquid-crystal phases, Ger. Offen. DE 3,612,238; *Chemical Abstracts*, **108**, 159109n.
12. Mitchell, A. R.; Pagoria, P. F.; Schmidt, R. D. (1997) Vicarious nucleophilic substitution using 4-amino-1,2,4-triazole, hydroxylamine or O-alkylhydroxyl-amine to prepare 1,3-

diamino-2,4,6-trinitrobenzene or 1,3,5-triamino-2,4,6-trinitro-benzene, *U.S. Patent 5,633,406.*
13. Mitchell, A. R.; Pagoria, P. F.; Schmidt, R. D. (1996) Vicarious nucleophilic substitution to prepare 1,3-diamino-2,4,6-trinitrobenzene or 1,3,5-triamino-2,4,6-trinitrobenzene, *U.S. Patent 5,569,783.*
14. Makosza, M. and Winiarski, J. (1987) Vicarious nucleophilic substitution of hydrogen, *Acc. Chem. Res.*, **20**, 282-289.
15. Pagoria, P. F.; Mitchell, A. R.; Schmidt, R. D. (1996) 1,1,1-Trimethylhydrazinium iodide: a novel, highly reactive reagent for aromatic amination *via* vicarious nucleophilic substitution of hydrogen, *J. Org. Chem.*, **61**, 2934-2935.
16. Coburn, M.D., Work in progress.
17. Meisenheimer, J. and Patzig, E. (1906) Directe Einführung von Aminogruppen in den Kern aromatischer Körper, *Ber.*, **39**, 2533-2542.
18. Seko, S. and Kawamura, N. (1996) Copper-catalyzed direct amination of nitro-benzenes with O-alkylhydroxylamines, *J. Org. Chem.*, **61**, 442-443.
19. Mitchell, A. R., Pagoria, P. F. and Schmidt, R. D. (2000) Amination of electrophilic aromatic compounds by vicarious nucleophilic substitution, *U.S. Patent* to issue.
20. Holleman, A. F. (1930) 1,3,4,5-Tetranitrobenzene, *Rec. trav. chim.*, **49**, 112-120.
21. Spencer, E. Y. and Wright, G. F. (1946) Preparation of Picramide, *Can. J. Research*, **24B**, 204-207.

SOME ASPECTS OF THE APPLICATION OF SMALL GRAIN POWDERS IN THE EMULSION EXPLOSIVES

P.KOHLICEK, E. JAKUBCEK *Department of Theory and Technology of Explosives, University of Pardubice, Pardubice, Czech Republic*
S. ZEMAN
Division of Commercial Explosives, ISTROCHEM, Inc., Bratislava, Slovak Republic

Abstract

An initiation ability, detonability and performance of the standard emulsion matrix, sensitised by perforated fine-grain powders, were studied. Initiation ability of the resulting mixtures depends first of all on the diameter and number of channels in the applied powders. On the basis of the relationship between performance and brisance of these mixtures it is stated that the sensitization in the above-mentioned sense leads to an approximately equal increase in the performance and to a distinctly lower increase in brisance as compared with the analogous application of PETN or RDX.

1. Introduction

Emulsion explosives of the „water-in-oil" type known under the code „EE" represent the last generation of development of industrial explosives. The last years have seen their considerable expansion, and thus they form (together with the ANFO explosives) the dominant assortment in the area of industrial explosives.

The most common way of sensitisation is introduction of hollow microspheres (MBs) based on silicates rendered hydrophobic or polymeric organic materials into the emulsion matrix. However, from the point of view of both the production and practical applications, this way of sensitisation cannot be considered as the optimum procedure, since the microspheres as the most expensive component markedly increase the price of those explosives, although they primarily determine the resistance of EE against hydrostatic pressure. From the viewpoint of explosion technology, they unfavourably affect the density of the explosive, and according to ref. [1] also partially the thermochemistry of the detonation transformation itself.

The emulsion explosives can also be sensitised by components of explosive nature such as trinitrotoluene (TNT), pentrit (PETN), and hexogen (RDX). However, the application of these sensitising additives results in partial elimination of some suitable properties of EE especially in the area of sensitivity to mechanic stimuli and physiologic activity of the explosive mass. In addition to it, the application of high explosives is economically uninteresting in this sense.

The reduction of military materials and partial disarmament in the states of former East Block have initiated research-development activities in the field of sensitisation of EE also with the use of delaborated smokeless gunpowders. In the context of these activities, e.g. researchers from Institute of Chemical Physics, Russian Academy of Sciences, have developed an explosive mixture [1,2] sensitised by a gunpowder which was not specified in any detail, and they present it as „a new type of emulsion explosives" giving a 10 - 20% increase in performance as compared with classic emulsion mixtures. The communication [3] by Kalacej *et al.* presents a mixture of this type (sensitised with pyroxilin powder) under the name Emulsen P. The density of this explosive is 1500 $kg.m^{-3}$, and the detonation velocity varies within the interval of 5200 - 5600 $m.s^{-1}$. The patent specification [4] by Adamec *et al.* deals with the dependence of detonation ability on the diameter of charge and the used smokeless powder (the patent claims involve the channel diameter in the applied powder).

In the context of research program of our Department we are studying the relationships between performance and brisance of emulsion explosives and/or the effect of some components of emulsion matrix on this relationship. The said research program also includes comparison of parameters of common commercial explosives (produced in the Czech and the Slovak Republics) with those of the EE sensitised by addition of fine-grain smokeless powders or, as the case may be, also high explosives. Preliminary results of the studies of emulsion explosives carried out in the above-mentioned sense are given in the present communication.

2. Experimental

2.1. COMPARISON SAMPLES OF EXPLOSIVES

The samples used for the comparison in the sense of dependencies in Figs. 1 and 2 included both current commercial products and explosives prepared on pilot-plant scale, in particular the following:

a) Powdered explosives: Polonit V (Istrochem Inc., Bratislava), Permon 10 and Permonex V 19 (both from Synthesia Division, Pardubice),

b) Gelatine explosives: Danubit 1 and Danubit 3 (containing 20 - 22% nitrate esters) and Danubit Geofex 2 (containing 40% nitrate esters) from Istrochem Inc., Bratislava, Perunit 22 and Perunit 28 (containing 22% and 28% nitrate esters, respectively) from Synthesis Division, Pardubice,

c) Plastic explosives: Semtex S 1A from Synthesis Division, Pardubice, and Composition C 4 from pilot-plant-scale production in our Department,

d) Classic emulsion explosives: Emsit M from Synthesis Division, Pardubice, Emulite 100, and 1 SO from Nitro Nobel AB, Impulsit and Emulgit 22 from Westpreng GmbH,

FRG, and mixtures produced on pilot-plant scale in our Department [S] denoted in Figs. 1 and 2 as I (a mixture of ammonium nitrate, water, fuel and emulsifier), II (a mixture of ammonium nitrate, sodium perchlorate, water, fuel and emulsifier), III (a mixture of ammonium nitrate, calcium nitrate, water, fuel and emulsifier), and IV (a mixture of ammonium nitrate, sodium nitrate, water, fuel and emulsifier),

e) Emulsion explosives based on standard matrix (see part 2.3) prepared on pilot-plant scale in our Department or in Istrochem Inc., Bratislava, sensitised and fortified by addition of fine-grain gunpowders (see part 2.2) or high explosives (the data were taken from Thesis [6]).

2.2. SAMPLES OF POWDERS USED

The applied samples of perforated fine-grain smokeless powders are characterised by the following data:

Sample	Type	Composition	$d_{channel}$ (mm)
1	Nc tp 3.0x 1.25/3.5 - KF 1	Nc, stabilisers, volatiles, graphite	0.28
2	S 070	Nc, stabilisers, water, graphite	0.25
3	Single-hole, tubular	Nc, stabilisers, graphite	0.25
4	6/7P - SBPFL	Nc, stabilisers, volatiles, graphite	0.15-0.20
S	Nc 7p 8.9x1.7/14(11/7)	Nc, stabiliser, volatiles	0.45

2.3. EMULSIONS WITH ADDED SMALL GRAIN POWDERS

All the experiments were based on the standard matrix described in Thesis [6]. Its composition is given in Table 1. This standard mixture is referred to as Matrix A (with oxygen balance 2,6 % O_2) in the following text.

TABLE 1. Composition of emulsion Matrix A

Compound	%
AN	61.8
CN	24
H O	7.1
Oil	3.5
Polymer	1.6
S 43	2

Note: AN ammonium nitrate
CN calcium nitrate
S 43 sorbitol sesquioleate

2.3.1. *Amount of Additive in Emulsion*

Attention has been focused on the optimisation of amount of delaborated gunpowder added as the sensitising additive of the emulsion matrix. For this purpose, the tubular gunpowder 1 with inner channel was used. The amounts of gunpowder 1 added gradually to the matrix were 20, 30, 40, and 50% by wt. The detonation velocity using the booster of 30 g Semtex S 1 A was measured on the samples with satisfactory consistency (20 %, 30 %, 40 %).

TABLE 2 Values of detonation velocities of EE sensitised with gunpowder

Powder content (% by wt.)	ρ (g.cm^{-3})	\multicolumn{3}{c}{D m.s^{-1}}		
		20 mm	30 mm	40 mm
20	1.463	-	-	1400
30	1.469	-	-	5400
40	1.472	1760	-	5600

2.3.2. *Tests of Charges of 32 mm Diameter*

In order to increase the sensitivity of the emulsion explosive sensitised with 30 % by wt. Powder 1, hollow micro-spheres from 3M company (80 μm) were added into the mass and carried out a test of detonation ability with detonator No. 8. The amounts of micro-spheres added to the explosive were 0.5, 1, and 1.5% by wt. The initiation power of detonator No. 8 was not sufficient for initiation of stable detonation of the samples modified in the way described.

The detonation velocity and performance of the charges of 32 mm diameter were measured using the booster of 30 g Semtex S 1A. The following amounts of gunpowder were introduced into the emulsion matrix: 1) 30% by wt. Powder 1; 2) 30% by wt. Powder 2; 3) 30% by wt. Powder 3.

The results are given in Table 3.

TABLE 3 Values of detonation velocities and performance

Sample	ρ (g.cm^{-3})	Performance (%)	D (m.s^{-1})
1	1.469	75.5	3830
2	1.464	74.5	4660
3	1.401	73.5	3920

2.3.3. *Adiabatic Compression*

The structure of powder 1 is characterised by an inner channel of 0.28 mm dimension. The passage of shock wave through the gaseous medium present in the channel results in compression of the gas and subsequent abrupt temperature increase with the consequence of more suitable propagation of exothermic reactions within the detonation transformation. However, the temperature increase in the channel depends on parameters of the medium (see Table 4).

TABLE 4 Temperature increase of gaseous medium depending on pressure in shock wave [7]

Medium (gas)	Pressure in shock wave (atm)	Temperature at shock wave front (°C)
Air	20	900
	100	3440
	200	5090
	500	7940
Argon	20	1480
	100	7480
	200	14980

In this sense, powder 1 was modified by introducing gaseous (liquid) medium into the inner channel to give:

1. Sample A - air in the channel
2. Sample B - water in the channel: the powder was evacuated with the help of water jet pump on a water bath
3. Sample C - argon in the channel: the powder was evacuated with the help of water jet pump and then kept in argon atmosphere 15 min.

Introduction of 30% by wt. powder 1 in the form of A, B, or C into the standard emulsion matrix provided the explosion mixtures referred to as samples A, B, or C, respectively, in Table 5.

The tests of detonation ability of the charges of 32 mm diameter were carried out with the use of booster of Semtex S 1 A, and the results are presented in Table 5.

TABLE 5 Results of tests of detonation ability

	mass of booster (g)					
Explosive with sample	0 (det.No.8)	20	30	40	50	60
A	-	-	+	4490, 2220 m.s^{-1} (the detonation extinguishes)	Not measured	Not measured
B	-	-	-	-	-	-
C	-	-		-	-	Not measured

2.3.4. Tests of Charges of 37 mm and Larger Diameters

Measurements of detonation velocities of charges with the diameters of 37, 40, 50, 60, and 65 mm, whose average weight was 1000 g, were carried out. For this purpose, explosives sensitised with various kinds of delaborated gunpowders, were prepared:

- matrix + 30% by wt. powder 1, p = 1.448 g.cm^{-3}
- matrix + 30% by wt. powder 4, p = 1.418 g.cm^{-3}
- matrix + 30% by wt. powder 5, p = 1.411 g.cm^{-3}
- matrix + 30% by wt. powder 5 + 1 % by wt. MB (microspheres from 3M company, 80 mm), ρ = 1.321 g.cm^{-3} (referred to as sample 6 in the Table)

TABLE 6 Results of measurement of detonation velocities

Sample No.	EE with powder	mass of EE (g)	mass of booster (g)	P (g.cm^3)	D (mm)	l (mm)	D (m.s^{-1})
1	1	-	30	1,448	37 PE		Incomplete detonation
2	1	-	30	1,448	50 PE		Incomplete detonation
3	5	500	100	1,411	Outer 50, inner 40, PP tube		
4	5	-	100	1,411	65 PE		Incomplete detonation
5	5	920	100	1,411	Outer 50, inner 40, PP tube		
6	5	920	100	1,411	Outer 50, inner 40, PP tube		
7	5	920	50	1,411	Outer 50, inner 40, PP tube		
8	1	900	100	1,448	Outer 50, inner 40, PP tube		Incomplete detonation
9	1	900	100	1,448	Outer 63, inner 60, PVC tube		Incomplete detonation
10	4	1000	200	1,448	Outer 63, inner 60, PVC tube		
11	6	-	50	1,321	Outer 50, inner 40, PP tube		

Note: PP polypropylene tube, PVC polyvinyl chloride tube, PE polyethylene sleeve

For initiation of charges, we used the gelatine explosive Danubit 2 (Istrochem Inc., Bratislava) with the following detonation parameters:

Oxygen balance	+1.5% O_2
Density	1.35 g.cm^3
Detonation velocity of packed charge	5000 m.s^{-1}
Performance	83%
Detonation transfer	6 cm

Performance values determined in samples No. 1, 10, and 11 were 84, 78, and 81 %, respectively.

2.4 PARAMETERS OF EMULSION EXPLOSIVES

2.4.1 *Performance of Explosives*
The performance values were determined by measurements in ballistic mortar [8] from the weight of the reference explosive (blasting gelatine), which causes the same deflection of the mortar arm as is achieved with 10 g of the explosive tested. For the measurement, the tested explosive was placed in a tin foil and brought to detonation with the help of detonator No. 8.

2.4.2. *Detonation Velocity*
The detonation velocity was determined by means of an electronic chronometer using a standard procedure consisting in measurement of the time needed for passing of detonation wave between two sensors placed at a definite distance. The sensors were placed inside the explosive [8]. The data of commercial explosives were taken from the producers' specifications.

2.4.3. *Dependence of Performance vs. Brisance of Emulsion explosives*
For the purpose of comparison of parameters of classic emulsion explosives (sensitisation with micro-spheres), emulsion explosives sensitised with the help of fine-grain gunpowders with those of common commercial explosives, the relationship between performance and brisance (expressed as product of density p and square of detonation velocity D^2) [5] was applied, see Figs. 1 and 2. Fig.1 includes the data of emulsion Matrix A sensitised with the fine-grain powders (data 1, 2, 4, and 6). Fig. 2 includes the data of emulsions sensitised with the help of high explosives PETN and RDX. These emulsions are based on Matrix B (a mixture of ammonium nitrate, calcium nitrate, glycol, water, fuel and emulsifier) with the oxygen balance of 0.8 % O_2, and from Matrix C (a mixture of ammonium nitrate, sodium nitrate, calcium nitrate, water, fuel and emulsifier) with the oxygen balance of 1.2 % O_2.

3. Discussion

The addition of 30% by wt. perforated fine-grain powder turned out to be the optimum sensitisation of the emulsion matrix used. An increase of this content up to 40% by wt. did not bring any significant increase in detonation velocity, density and initiation ability of the resulting explosion mixture (see Table 2). A content of 20% by wt. powder in emulsion explosive did not ensure a stable course of its detonation. The amount of 30% by wt. of the above-mentioned sensitizer in this explosive represents the optimum from the point of view of its consistency too.

The results of detonation tests of emulsion explosives with the optimum content of additives placed in polyethylene sleeves indicate that the charge diameter of 32 mm is

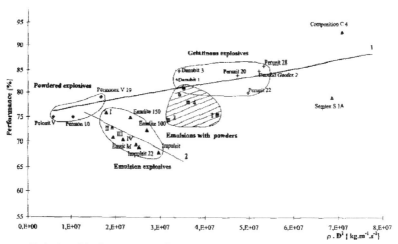

Figure 1. Relationship between performance and brisance for emulsions with powders

near the lower critical diameter of these explosives. The values of both detonation velocity and performance indicate that the detonation is not completely stable here (the charges can partially deflagrate). The sensitivity of these explosives to initiation with detonator No. 8 for the charge diameter of 32 mm could not be increased, not even by addition of hollow micro-spheres of 80 pm diameter in the amount of 1.5% by wt. (see Table 3).

The experiments focused on affecting the sensitivity to initiation of emulsion explosives (charges of 32 mm diameter) by means of media closed in channel of powder 1 gave the results presented in Table 5: if the medium was air or water, then the results roughly corresponded the expectation. However, negative results were obtained in the case of argon used as the medium, which contradicts the data on compression heating of this gas (see Table 4). This discrepancy can be interpreted as follows: the initiation and primary propagation of detonation are mostly realised in pores and spots of non-homogeneity in the explosive whose diameters are 4 - 20 μm (at current initiation pressure), whereas a gaseous medium in the powder channel contributes to the propagation of detonation only after its stabilisation [9] (after its velocity is stabilised). Hence, for the initiation of detonation it is important to have a gaseous medium present in the form of occluded micro-bubbles at the surface or in pores (spots of non-homogeneity) of the powder. When introducing the powder into emulsion matrix at normal conditions, the formation of these occluded air micro-bubbles can be expected (sample A in Table 5). The powder used in preparation of sample C was rid of air in vacuum and subsequently left in argon atmosphere 15 min. This time, obviously, was insufficient for perfect saturation of pores and spots of non-homogeneity of powder with the said gas, which made itself, felt in unsatisfactory sensitivity of the corresponding explosive to initiation.

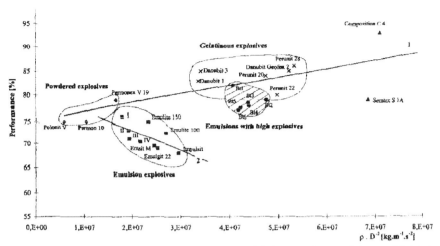

Figure 2. Relationship between performance and brisance for emulsions with high explosives

Bt1 – C, Petn 30 % by wt., **Bt2** – A, Petn 30 % by wt., **Bt3** – B, MB 2 % by wt., RDX 30 % by wt.
Bt4 – B,Petn 30% by wt., **Bt5** – C, MB 2% by wt., RDX 30% by wt., **Bt6** – A,. MB 2% by wt., RDX 30% by wt.,

The results given in Table 6 show that the detonation ability of emulsion matrix with added 30% by wt. powder (the charge diameters from 40 mm to 65 mm) predominantly depend on the type of the powder used (i.e. on the diameter and number of channels in its grain) and also on the strength of the charge container. The best results were achieved by application of multiperf powder (7 holes) 5 with the channel diameter of 0.45 mm (the detonation velocities of 5700 - 5800 m.s^{-1}), worse results being obtained with the application of multiperf powder (7 holes) 4 with channel diameters of 0.15 - 0.20 mm (the detonation velocity of 5100 m.s^{-1}). An addition of 1% by wt. hollow micro-spheres to the emulsion explosive with the content of powder 5 results in a density decrease of the final mixture and lowering of detonation velocity to 5200 m.s^{-1}, the sensitivity to initiation being not significantly increased (see the comparison of booster mass of the samples 7 and 11 in Table 6); however, the performance and brisance of this mixture approach those of gelatine explosives (see Fig. 1). The explosive mixture sensitised with powder 1 (one channel of 0.28 mm diameter) detonated incompletely.

The specification of explosives performance with the help of ballistic mortar forms a current part of output checking of explosives in firms producing commercial explosives in both the Czech and the Slovak Republics. Due to construction features of this mortar (perfect closing of sample of 10 g explosive by a heavy projectile) and the initiation power of detonator No. 8, the conditions of this test markedly increase the initiation ability and detonation ability of the emulsion explosives studied.

The relationship between performance and brisance of the explosives studied (compared with an analogous relationship of current commercial explosives) is presented in Figs. 1 and 2; in these Figs., the brisance is expressed by the product of density ρ and square of detonation velocity D^2. In the case of commercial explosives, an increasing content of explosive sensitizer roughly corresponds to increases of both performance and brisance of these products (see the straight line 1 in Figs. 1 and 2). The lowest values, in this sense, are those shown by ammonium nitrate powdery explosives (Amatol type). The emulsion explosives composed only of ammonium nitrate solution, fuel phase, and hollow micro-spheres (mixture I in Figs. 1 and 2) are comparable with the powdered ammonium nitrate explosives of Amatol type both in performance and brisance (the data of 12 mixture I correlate well with the straight line 1); also the donor-acceptor sensitivity of emulsion explosives is practically identical with the corresponding sensitivity of ammonium nitrate explosives containing explosive sensitizers [10].

Incorporation of oxidising agents based on nitrates or perchlorates of alkali metals or alkali earth metals (SN - sodium nitrate, CN - calcium nitrate, PC - potassium perchlorate - mixtures II, III, and IV in Figs. 1 and 2) into the discontinuous phase of emulsion matrix leads (even if the oxygen balance of the final mixture is maintained) to a drop in performance and, at the same time, to an increase in the brisance of final explosive (straight line 2 in Figs. 1 and 2) [11]. The explosives with explosive sensitizer (the data of straight line 1 in Figs. 1 and 2) show no such effect.

The sensitisation and fortification of emulsion matrix with brisant explosives PETN and RDX in an amount of 30% by wt. leads logically to increases in both performance and brisance of the resulting mixtures, which thus approach very much to those of gelatine explosives (see Fig. 2). Also here it is possible to see a distinct effect of the presence of nitrates of alkali metals or alkali earth metals on the performance of the final emulsion explosive (see Fig. 2). Incorporation of perforated fine-grain powders into the emulsion matrix also in an amount of 30% by wt. leads to an approximately equal increase in performance but to a distinctly smaller increase in brisance of the resulting explosive, as it is the case with the same application of PETN or RDX (compare Fig.1 and Fig. 2).

4. Conclusion

The sensitisation and fortification of emulsion matrix A with oxygen balance 2.6 % O_2 by means of the perforated fine-grain powders in the amounts of 30% by wt. leads to an emulsion explosive characterised by the maximum detonation velocity and density along with attaining the optimum consistency of this explosive. Its initiation ability depends first of all on the diameter and number of channels in grain of the applied perforated powder (the most suitable is multiperf 7 holes powder with channels of 0,45 mm diameter). The simultaneous incorporation of hollow microspheres into the emulsion explosive leads to a drop in density and detonation velocity of the resulting mixture without any distinct increase of its sensitivity to booster; the performance and brisance of such mixture can approach those of some gelatinous explosives. The increase in initiation ability of these explosives, by means of argon as the gaseous medium in the channels of the applied powder turns out to be impassable. The sensitisation and fortification of emulsion matrix with perforated fine-grain powders leads to an approximately equal increase in the performance and to a distinctly lower increase in brisance of the resulting explosive as compared with the application of PETN and RDX in the same sense and amount.

Acknowledgement
The authors are indebted the management of Division of Commercial Explosives, ISTROCHEM, Ltd., Bratislava, for its kind interest in and support of realization of this study. The authors wish also to thank Assoc. Professor Boris Vetlicky and Assoc. Professor Pavel Vavra, Department of Theory & Technology of Explosives, Univ. of Pardubice, for helpful discussions.

5. References

1. V. V. Odintsov, V. I. Pepekin and B. N. Kutuzov: (1994) Evaluation of Detonation Parameters of the New Kind of Emulsion Explosives, *Khim. Fiz.* **13**(12) 131,
2. V. V. Odintsov and V. I. Pepekin: (1995).(Comparable Characteristics of some Kinds of Commercial Emulsion Explosives, *Khim. Fiz.* **14**(7) 132
3. V. I. Kalatsev, B. V. Matseevich, V. P. Glinskii, N. I. Plekhanov, N. K. Shaligin, O. F. Mardasov and A. G. Fridman: (1995) Explosive Materials from Utilized Ammunition: Problems, Solutions, Assortment, *Bezopasnost' truda v promyshlennosti* No. 12, 32
4. Yu. D. Adamets, V. E. Annikov, S. N. Puzyrev, B. V. Matseevich, V. P. Glinskii, V. I. Kalatsev N. K. Shaligyn, N. I. Plekhanov and O. F. Mardasov: An Emulsion Explosive Composition*). Rus. Pat. RU 2092473* (Oct. 10`h, 1997).
5. P. Kohlicek: Comparison of Parameters of Emulsion and Common Commercial Explosives.(1999*) Proc. of the 2nd Seminar „New Trends in Research of Energetic Materials"*, April 1999, Univ. of Pardubice, p. 159.
6. P. Kohlicek: Emulsion Explosives, II. (1998), *Thesis, Univ. of Pardubice*, June 1998.
7. F. Vokac: Explosive Argon Flashes, (1966). *Prumyslova chemie* **5**(11) 3
8. *Notice of Czech Mining Authority* No. 246/1996 of Law Collect, establishing more detailed conditions for allowing explosives, explosive objects and aids into use, and their testing.
9. R. Belmans and J.-P. Plotard: (1995), Physical Origin of Hot Spots in Pressed Explosive Composition. *Journal de Physique IV, Coll. C4, suppl. au Journal de Phys.* III, 5, C4-61
10. J. Denkstein and J. Jarosz: (1992), Some Important Properties of Emulsion

Explosives, *Proc. of the Conf. Production and Utilization of Commercial Explosives*, Dum techniky CSVTS, Ostrava, pp. 26-36

11. P. Kohlicek and K. Bezkocka: (1999), Relationship between Performance and Brisance of Emulsion Explosives, *Proc. of the Int. Conf. "Blasting Technique 1999"*, High Tatras, May 1999, p. 49.

6. Supplement

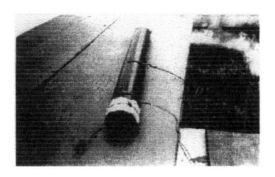

Figure 3. Measuring of detonation velocity of cartridge with diameter 40 mm (emulsion with powder I)

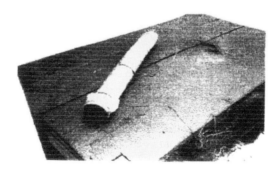

Figure 4. Measuring of detonation velocity of cartridge with diameter 6 Mm (emulsion with powder 1)

Figure 5. Incompletely detonated cartridge of emulsion containing powder 5 with diameter 65 mm.

JOINT DEMILITARIZATION INTEGRATION
Demilitarization Research and Development Program Integration in the United States

J. Q. WHEELER
U.S. Army Defense Ammunition Center, McAlester, OK,
J. LIPKIN, Ph.D.
Sandia National Laboratories, Livermore, CA

1. Introduction

Historically, standard techniques of removal, disassembly, incineration or open burning/open detonation (OB/OD) were viewed as both safe and efficient. As environmental awareness increased and potential health and safety risks became known, a requirement for alternative destructive technologies and enhanced resource recovery processes began to emerge. A great deal of effort has been expended to improve the acceptability of demilitarization processes and to propose and implement alternatives for specific items. However, since demilitarization has not been a major component of munition development and acquisition life cycle, there have been few opportunities to support technology development and implementation. Most work has been done as process improvements or operations support under the U.S. Army's Ammunition Peculiar Equipment Program or as industrial plant equipment. Since Fiscal Year (FY) 90, support has come from the Joint Service Large Rocket Motor Demilitarization Program; the Joint Department of Defense (DoD)/Department of Energy (DOE) Munitions Technology Development Program, Technology Coordination Group (TCG) IX on Munitions Life-cycle; the U.S. Army's Conventional Ammunition Demilitarization Research and Development (R&D) Program; and the Navy's Ordnance Reclamation Program. Support has also been received from environmental science and technology funds for the U.S. Military Services, the Strategic Environmental Research and Development Program (SERDP), and the Environmental Security Technology Certification Program (ESTCP).

2. Requirements

The demilitarization technology research and development program is a cooperative interservice, interagency effort with industry, academia, and international participation dedicated to the development of safe, efficient and environmentally acceptable processes for the resource recovery and recycling (R3) or disposition of strategic, tactical, and conventional munitions including explosives, and rocket motors. Efforts in the demilitarization technology R&D program emphasize environmentally compliant technologies to enhance existing methods for munitions. There are currently over 500,000 tons of these materials requiring disposition with an additional 700,000 tons forecast to pass through the stockpile during the next 10 years. Collectively these figures represent over one third of our current active inventory. Located in 27 states, the Pacific, and Europe, the stocks are stored in facilities with a mix of capabilities unique to each site. These facilities tend to be extremely diverse in design and function and in many cases they are unique in the demilitarization community. Additionally, the demilitarization program has begun to assist in the support of non-stockpile items, such as those buried at closing installations or during operations for clean-up of Formerly Used Defense Sites (FUDS).

To characterize the stockpile, the Military Services use the Munitions Items Disposition Action System (MIDAS) program. The MIDAS program forms the bridge between technology research and development and the demilitarization user community. MIDAS provides program execution support and technical assistance, however, the primary thrust of the program is to identify and characterize munitions items in direct support of resource recovery and recycling operations.

The MIDAS Program was established in FY 93 at the U.S. Army Defense Ammunition Center (DAC), McAlester, Oklahoma. The purpose of the program is to identify alternatives to open burning/open detonation (OB/OD), and to provide a systematic approach for disposition of unwanted munition items. Significant progress has been made in development of the program that supports demilitarization execution, resource recovery and recycling (R3), R&D technology application, and environmental permitting.

The central core of the MIDAS Program is the identification (characterization) of munitions components and constituents. Characterization of munitions is performed and managed by personnel at DAC with munition plants and/or depots assisting in the effort as required. The characterization data is then input into the MIDAS database designed by Argonne National Laboratory and centrally located at DAC. Government, Industry, and Academia have access to MIDAS data.

The program features easy to use menus and searches to view or print hierarchical listings of munition components, parts, materials, and bulk items (e.g., surface finishes and plating);

including material descriptions, specifications and weights. Twelve standard reports are provided to include detailed structure; constituents of energetic materials; and Resource Conservation and Recovery Act (RCRA) Regulated Material Reports. Other features include scanned images of munitions, resource recovery and disposition (RRD) inventory searches, and component/part/material usage searches. The data has been used to support numerous demilitarization contract solicitations, installation demilitarization projects, and State environmental permit applications.

The rational for pursuing demilitarization technologies is driven by a variety of issues. For example, there are over 700 energetic fillers in the current RRD inventory. The geographic dispersion and installation specific capabilities within the safety and environmental regulatory framework also drive requirements as does the simple need to provide operators with modern safe, efficient, and economic processes. In the 1980's and the 1990's agreements/treaties or the possibility of agreements/treaties, such as, Intermediate-Range Nuclear Forces, Conventional Forces Europe, START, START II and the international movement to ban anti-personnel landmines have also impacted technology requirements.

Stockpile stratification and analysis have been performed by the individual Military Services and form the basis for requirements generated by the Joint Ordnance Commander's Group (JOCG). The JOCG is a Joint Service organization consisting of flag officers who operate the U.S. munitions base. To support their analysis the Services were guided by overarching principles to include the best strategy to minimize the stockpile while always considering environmental and economic factors. In short, their objectives are to reutilize excess serviceable stocks whenever possible, to reduce inventory, maintenance, surveillance and demilitarization cost, and to free up storage space.

3. Demilitarization Technology Research and Development

Demilitarization technology R&D was initially focused on large rocket motors containing hazard class (HC) 1.3 (ammonia perchlorate) or HC 1.1 (RDX/HMX based) propellant. These projects were focused by user requirements based on safety and environmental issues, stockpile challenges, or opportunities to advance the demilitarization base capabilities. The demilitarization technology development strategy included near-term pay off projects, continuing investments, and long-term projects. Demilitarization technology development is firmly linked to the U.S. DoD Science and Technology Strategy in that it 1) strengthens links to customers, 2) builds upon and complements component Strategic Planning, 3) adds value by horizontal integration of corporate research/technology and 4) utilizes a team effort- Government, Industry and Academia.

3.1. JOINT DEMILITARIZATION TECHNOLOGY PROGRAM (JDTP)

The U.S. DoD, in 1997 directed establishment of the JDTP to develop safe, efficient, environmentally acceptable demilitarization processes for R3 or disposition of conventional ammunition, rocket motors, and energetics. The JDTP is a cooperative interservice, interagency effort focused as the sole DoD program dedicated to the development of demilitarization technology.

The JDTP focus is on 1) Transitioning Technology to address warfighting needs, 2) Reducing Costs, 3) Strengthening the Industrial Base, 4) Promoting Basic research, 5) Assuring Quality. The program assures that joint needs are met, extensive commercial involvement, and that the demilitarization industrial base achieves a cost effective technology transfer through a cooperative conduit of teamwork and partnership. It takes commercial processes and applies them to military applications. Building on DoD strengths, the program is scoped globally. An annual Global Demilitarization Symposium serves as an international demilitarization program review and a website is maintained by the MIDAS, which incorporates the JDTP.

The JDTP plan represents near, mid, and long-term goals of the demilitarization community. Annual assessments and adjustments will occur to ensure the DoD maintains a coherent technology development strategy through the Future Years Defense Plan. The program is executed by the Joint Services with DOE, academia, and industry partnerships. The goal of each project will be transition into the government and commercial demilitarization base to enhance safety, efficiency, and environmental acceptability in a manner that complies with all applicable environmental laws.

3.2. U. S. ARMY CONVENTIONAL AMMUNITION DEMILITARIZATION TECHNOLOGY RESEARCH AND DEVELOPMENT PROGRAM

This program supports a continuing technology evaluation of demilitarization methods for existing conventional ammunition and conventional ammunition recovered from FUDS. This program leverages funding from other programs to complete the development and demonstration of new, safe, and environmentally acceptable alternatives to OB/OD for R3 equipment and processes to reduce the extremely large inventory of munitions in the Army's resource recovery and disposition account and recovered munitions from FUDS.

3.3. JOINT DOD/DOE MUNITIONS DEMILITARIZATION

The U.S. DoD and U.S. DOE, through a memorandum of understanding, have agreed to use the unique facilities and capabilities of the U.S. DOE laboratories to explore and apply advanced technologies of mutual interest and identified importance to the U.S. Armed

Services. This program exploits the extensive and highly developed technology base resident in the National Laboratories relevant to achieving the goal of developing applicable, cost-effective conventional munitions, and leverages the U.S. DoD investments with matching U.S. DOE investments. The current program supports projects in warhead technology, energetics, advanced initiation and fuze development, munitions demilitarization, service life technology, and computer simulation. A specific DoD Service laboratory sponsors each of these active projects. Under the preview of the TCG IX, numerous programs have been leveraged including, but not limited to, molten salt destruction, chemical conversion of explosives and propellants, emission monitoring, cryocycling, cofiring of desensitized energetics in commercial boilers, laser cutting, robotic applications, propellant stability analysis, and base hydrolysis.

4. Conclusion

If not used for their intended purpose, strategic, tactical and conventional weapons systems whether excess, obsolete or unserviceable, currently fielded, or being developed for future applications, will require disposition planning and capability. To address these items the U.S. Congress and the U.S. DoD have recognized the need for demilitarization technology development. The JOCG Munitions Demilitarization and Disposal Subgroup directs the JDTP in a partnership that coordinates and leverages resources from the Army's Conventional Ammunition Demilitarization Technology R&D, Joint DoD/DOE Munitions Demilitarization TCG IX, SERDP and ESTCP. This has resulted in partnering for development of technologies that address specific requirements in a safe, efficient, and environmentally acceptable way. As technologies mature and regulations are finalized, the requirements will be addressed by transitioning the technologies into the U.S. DoD and commercial base. The annual Global Demilitarization Symposium and Exhibition features presentations representing all the overarching technologies under development for these programs. For More information contact Mr. James Q. Wheeler at the US Army Defense Ammunition Center, McAlester, Oklahoma 74501-9053, USA; phone (918) 420 8901, or visit the DAC Webpage at http://www.dac.army.mil

NEW ASPECTS OF THE DEVELOPMENT OF THE INDUSTRIAL EXPLOSIVES AND MATERIALS ON THE BASIS OF UTILIZED POWDERS

N.G.IBRAGIMOV. E.F.OKHRIMENKO, E.H.AFIATULLOV,
U.M YUKOV, U.M.IVANOV, I.H.GARAEV
State Unitary Enterprise Research Institute of Polymer Materials
Perm, Russia

Nowadays great amount of powders and solid rocket propellants was accumulated which are necessary to utilize because of termination of period of their serviceability and moral obsolescence.

There are two methods of utilization of powders and propellants. This first and the simplest method are incineration (burning). However the incineration requires organization of special stands with complex system of purification (clearing) of the combustion products from the toxic compositions and considerable material expenditure.

Other method of utilization is the production of the civil purpose products from powders and propellants Use of powders and propellants for the production of articles of civil purpose gives not only economic effect, but also allows at modern reduction of output of powders to conserve production potential - know-how and personnel since the utilization of powders, as a rule, is carried out at the same plant, where they were produced.

In Research Institute of Polymer Materials (NIIPM) the intensive investigations (studies) on use of energy of combustion and explosion both powders and propellant in national economy are carried out. The main directions of studies on use of the utilized powders and propellants are given below and are shown in Fig. 1.

1. Production of water stable explosives for mining.
2. Production of water-emulsion and water-gel industrial explosives.
3. Production of pressure accumulators for intensification of the wells at oil production.
4. Production of signal and tire lights of different colours. 5. Production of paintwork materials from pyropowders.

The main questions which are actual to the present time had to be solved when producing water stable explosives from powders and ballistite-type propellants.

a) Providing of detonating ability of powders with the high level of critical diameter of their detonation
b) Grinding of artillery tubes

c) Grinding (preparation) of grains-blanks and solid propellants
d) Grinding of pyropowders.

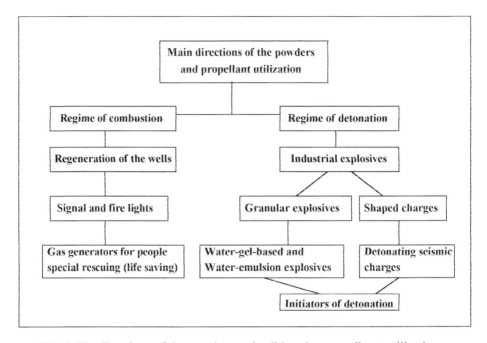

FIG. 1. The directions of the powders and solid rocket propellants utilisation

With the purpose of decrease of critical diameter of detonation of the utilized ballistite-type powders and propellants the studies on searching of the effective sensitizes providing simultaneously ecological purity of explosion products of the explosives were carried out. As a result of these studies series of the components were found, which one could reduce critical diameter of detonation of powders and propellants in 2.0 - 4.0 times and simultaneously ensure ecological purity of the explosion products. There are no heavy metal salts: lead, nickel, etc. in explosion products.
The unit for grinding of artillery tubes from ballistite-type powder is based on the equipment of ballistite-type powder production and wastes processing - on machines such as "Kruppa", rollers such as "Bolshevik".

For cutting and grinding of large-sized propellants special machines (tools) with output till 500 kg /hour are developed. The construction of a tool for preparation and grinding of solid rocket propellants with f diameter of 250 mm with the output up to 900 kg /hour and energy consumption approximately 20 kW per ton with the use of an original engineering solution is proposed.

A powder grinding is one of the complicated problem in the process of their utilization, which requires the solution.

The complexity of the powder grinding is that it is in glassy state and equipment used when cutting thermoplastic ballistite - type powders and propellants are not suitable for their cutting.

In this connection we have developed the original tool (machine) for cutting of tubes from powders into granules of any given length with output 500 kg/h at energy consumption 4,0 kW per ton. Such grinding of tubular powders allows to receive explosives from them with high bulk weight in combination with ballistite - and other powder types.

On the basis of the performed studies on searching of the effective sensitizers, devices for cutting and grinding, improvement of know-how of the production from utilized ballistite-type powders, solid rocket propellants, pyropowders, water stable industrial explosive (dibazit) and diammozit for the dry wells were developed. The main explosive and physic-chemical characteristics of the new industrial explosives in comparison with regular one are shown in tab.1.

TABLE 1 The main explosive and physic-chemical characteristics of explosives

Characteristics	Dibazit	Diammozit	Granulotol	Grammonit
1. Water resistance, not less 24 hours	60,0	-	30, 0	0,5
2. Detonation velocity, km/s				
- in water	5,8 – 6,0	-	5,0 – 5,5	5,5 – 6,0
- in dry state	2,8 – 3,6	2,6 – 3,2	4,0 – 4,6	4,0 – 1,7
3. Critical diameter of detonation, mm				
- in water	25 – 40	-	25 – 30	10 – 15
- in dry state	100 – 120	100 – 150	60 – 80	40 – 60
4. Oxygen balance, %	-25 to -45	0 to –15	-74	-45,9
5. Heat of explosion, kcal/kg	800 - 900	920 – 1100	980	940
6. Volume of gases, L / kg	850 – 900	900 – 930	750	800
7. Impact sensitivity, GOST 4546, %	8 – 12	12 – 24	-	12 – 16
8. Strength, GOST 4546, cm3	280 – 300	340 – 380	285 - 295	330 – 340
9. Bulk density, g/cm3	0,8 – 0,9	0,85 – 0,95	0,9 – 1,0	0,85 – 0,9

When producing water-emulsion and water-gel industrial explosives additional grinding of granules of ballistite type powders, solid rocket propellants, pyre powders into crumb up to the sizes of particles 0,5-0,8 mm with the help of special disk mills, multistage cutting machines such as PC - 250, PC - 500 is necessary. Ballistite-type powder, solid rocket propellant, pyre powder crumb with the particle sizes of 0,5-0,8 mm are suitable for production of water-emulsion and water gel industrial explosives.

Using the above-stated tools (machines) and mills in PMRI the know-how of 0,5-0,8 mm and compositions of water-gel and water-emulsion explosives were developed. The main characteristics and compositions are given in a tab.2.

TABLE 2. Composition and the main characteristics of water-emulsion and water-gel explosive

Composition, % mass	Aquanitrate	Emulpor
1. Oxidant (ammonium, sodium, potassium nitrates)	50 – 63	50 – 63
2. Fliel (ballistite-type powder, solid rocket propellant, pyropowder	20 – 30 12 – 16	20 – 30
3. Water	4 – 6	12 – 14
4. Antifreeze	0,8 – 0,15	-
5. Thickening agent	0,05 – 0,1	-
6. Cross-linking agent	-	-
7. Mineral oil	0,1	3 – 5
8. Gas- generating components		-
Characteristics	**Characteristics meanings**	
1. Heat of explosion, kcal/kg	880	920
2. Critical diameter of detonation, mm	65	45
3. Detonation velocity, km/s	3,8 – 4,2	4,2 – 5,2
4. TNT equivalent	0,7	0,72
5. Charging density, g/cm	1,2	1,3

The next direction of PMR Study of the utilized powders and propellants application is the development of the universal shaped charges intended for explosive cutting of metal construction, concrete, reinforced concrete, rock, wood (timber) and so on. The different standard sizes of universal shaped charges are designed, technical characteristic of which are given in tab.3.

TABLE 3 Technical characteristics of the utilized shaped charges

Type of the charge	Mass without lining, g/m	Mass with lining, g/m	Thickness of the cutting barrier, mm
YKZ1-O1	1300	1800	12 - 14
YKZ 1-02	2600	3200	15 - 17
YKZ 1-031	45002	60003	20 - 254
YKZ 3-O 1	75	110	3 - 4
YKZ 3-02	120	160	6 - 8
YKZ 3-06	2600	3900	30 - 33

Studies on the development of the effective gas generators, pressure accumulators with the use of utilized powders and propellants, providing increase of production rate of non-commercial oil wells are also carried out (performed). Nowadays processing of (out of date) powders and propellants into effective pressure accumulators of oil wells is carried out according to the following scheme: grinding of powders ant propellants, adding to the preliminary grounded powder pellets a granule of the components (sensitises, filling agents) during rolling as much as possible lowering their burning rate, further moulding of non perforated end - burning grains and their restriction along external surface. The wells pressure accumulators manufactured by the above-mentioned method provide long-term process of burning of charges (10-30 min on depth of 1,5-2,0 km). The incineration of such charges in oil wells gives considerable effect on barothermo-gasochemical attack on oil layer. The wells oil content after treating by the end-burning charges of long action manufactured from utilized powders increases in 3,0 - 6,0 times in comparison with the wells oil content before treating.

The interesting studies on fabrication of signal and firelights with the use of utilized ballistite type powers and propellants were performed. As a result of these studies the formulation was produced and know-how of fabrication of signal and firelights of five different colours: red, green, yellow, blue, white is developed. The formulation of signal and fire lights consists of 60-80 % of mass utilized BP and solid rocket propellant, 10-15% of mass - metal powders or metal alloys as energetic ingredient and 15 - 20 % of mass of special components as inorganic salts of different metals reshaping (forming) colour and light radial spectrum.

For production of signal lights regular semicontinuous "Know-how" of ballistite-type powders, allowing fabrication of elements of any sizes with output not less then 1000 pieces in hour is used.

In addition to usage of out-of date (obsolete) pyro powders in industrial explosives the method of their clearing and decrease of viscosity is found and also the know-how of use of pyro powders in production of paintwork materials is developed.

The most complicated technical solution in a given direction is clearing of pyro powders from the content of diphenylamine. The presence of diphenylamine in pyro powders sharply reduces the quality of paintwork materials manufactured produced with applying of utilized PP. When storing and affecting of light the film from such paintwork materials changes its colour, which is extremely undesirable.

For exception of the specified phenomena the original methods of PP clearing from diphenylamine are found by chemical binding it with the other components and by removing from the PP composition. The technological receptions of viscosity decrease of PP with high nitrogen content for their further use in production of fast drying paintwork materials, paint primers, enamels, etc. are found.

Conclusion

The formulation and know-how (technology) of the industrial explosives Dibazit and Diammozit fabrication using utilized ballistite-type powders and propellants with maximum use of the existing equipment for ballistite-type powders production are developed.

The machine (tool) for cutting of pyro tubular powders on any given length with output of 500 kg/hrs at energy consumption 4,0 kW per ton is designed.

The technological receptions of the utilized powders and propellants grinding in chips with the particle size 0,5 - 0,8 mm are found.

The know-how of the grains- detonator fabrication and detonating seismic charges of any sizes from the utilized powders with the wide spectrum (range) of power of the detonating wave is developed.

Methods of burning rate decrease of utilized powders at the expense of inhibitors of combustion, filling agents adding to their composition are found. Using low-burning powders (U=3-4 mm/s, P= 10 MPa) the effective accumulators of pressure with operation time 15-30 min in the wells on depth of 1500-2000 m are produced.

Using of such charges rises production rate of unprofitable oil wells in 3-6 times.

The know-how of universal shaped charges fabrication using the utilized BP and solid rocket propellants for cutting and welding of metals for reinforced concrete, concrete and other constructions cutting is developed.

The formulation and know-how of the signal and fire lights of a wide spectrum of their glow using utilized powders and propellants are developed.

The methods of clearing of pyro powders from diphenylamine, methods of decrease viscosity of pyro powders are found and the know-how of pyro powders usage in production of paintwork materials is developed.

On all aspects (kinds) of products of civil purpose using utilized powders and propellants technical and technological documentation are developed. The resolutions of State municipal technical Inspection on their application in industrial conditions are obtained.

SMALL-SIZED SOURCES OF SEISMIC SIGNALS ON THE BASIS OF HIGH ENERGY CONDENSED SYSTEMS FOR TRANSITION AREAS GROUND - WATER AND ALMOST INACCESSIBLE REGIONS OF A LAND

N.M. PELYKH *, N.M.PIVKIN *, A.P.TALALAEV *, R.P.SAVELOV **, Y.A.BYAKOV ***
*State Unitary Enterprise Research Institute of Polymeric Materials, Perm, Russia,
** Perm State University, Perm, Russia,
*** State Enterprise NIPIokeangeophyzica, Gelendhzik, Russia

For conducting seismic studies the sources of seismic signals (SSS) of vibratory and impulse excitation are required. At vibratory excitation the variable force acting during the time not exceeding 5 ... 10 s is continuously applied to the spots of media. The generated seismic signals depending on the features of SSS are usually found in frequency range from 5 up to 250 Hz. At impulse excitation the time of the stimulating force action is less than the period of the excited vibrations, but sufficient to give spots of media velocity different from zero point.

Existing SSS safely work in many places. However for transition areas ground water, almost inaccessible regions of a land, and also for some other regions their usage, are undesirable or impossible. In this connection the necessity of the development of SSS of principally new type is obvious. Such SSS can be developed in devices on the basis of high energy condensed systems (HECS).

In connection with the conversion of the plants of special engineering chemistry and necessity of utilization of HECS the development of SSS on their basis has begun. The studies included firing bench tests of pilot samples of SSS, and after their improvement - experimental - methodical seismic explorations.

The small model rocket motors such as TRS-45, and on the basis of the camera B-1 were selected. The charges were made from model double-based and composite solid propellants. Their mass was changed from 50 up to 330 g.

The firing tests were carried out in air and in special vessel with water on the depth of 0,5 m.
In the fig. 1 and 2 two of several pilot samples of SSS of impulse and vibratory excitation testing on the stand are shown. As examples, in the fig. 3 and 4 the regimes of combustion of HECS obtained with their help are shown.

As a result of the bench tests there were induced, recorded and studied the regimes of operation of the devices which presented the interest for seismic explorations. For the impulse regime the times from 18 up to 25 ms, and impulse of force up to G00 N s were reached. The vibratory regime was characterized by pressure oscillations with frequencies up to several tens hertz with amplitudes reaching several megapascals and duration about several seconds.

During experimental - methodical seismic studies under water and in a well, filled with water, 30 m depth the pilot samples of SSS, working in impulse regimes were tested. When working under water hydrophones and recording instrumentation of Brul and Kjer firm were used. Data reproduction was carried out with the help of analyser ONO-SOKKI CF-930. In a fig. 5 oscillations of a sound of one of trials confirming a key opportunity of the elastic waves induction in water at the expense of vibratory combustion of HECS are shown.

For generation of the elastic waves from the surface of the ground the pilot sample of SSS of impulse excitation housed in a tube on a plate was used. During combustion of a charge there was recoil in a ground through this plate. On the seismic profile of 2,5 kms length for comparison the operations with this device, and also with an explosive SSS, used in seismic studies, were executed. During the trials of new SSS it was clarified, that on separate parts of the seismograms a proportion the signal / clutter was high enough and did not exceed the wave field for the explosive SSS.

In a fig. 6 the part of a depth cross-section for the tested ground SSS, obtained by the results of decryption of the seismograms, was shown. The reflective horizons on depths up to 1,8 kms were well scanned.

CONCLUSIONS

Key opportunity of excitation of the elastic waves in different media with the help of the devices containing high energy condensed systems, working in vibratory and impulsive regimes was firstly shown.

The proposed devices can find wide application as portable sources of seismic signals especially when the usage of the existing one is undesirable or impossible.

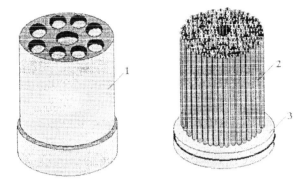

Figure 1   Pilot sample for land seismic exploration
1 -   body with nozzle ports
2 -   charge (double-base propellant)
3 -   lid with aperture for gauge pressure

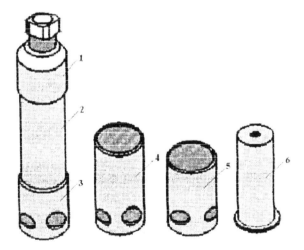

Figure 2. Pilot sample for underwater and well seismic exploration

1 - block with aperture for gauge pressure
2 - camera
3 ... 5 - nozzles with peripheral ports
6 - charge (composite propellant)

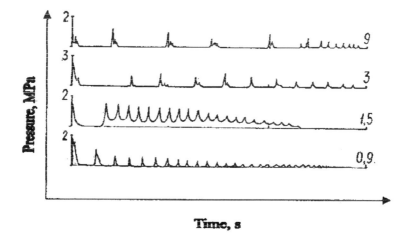

Figure 3. Vibratory burning for different kinds of HECS

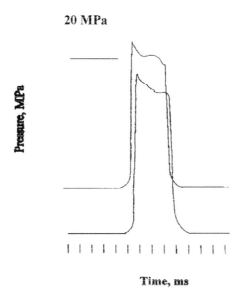

Figure 4. Example of operation of two impulse motors
Interval of time - 10 ms.

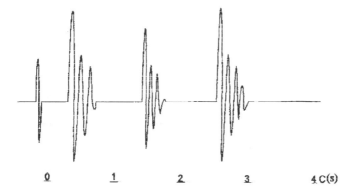

Figure 5. Fragments of underwater vibratory burning after start
Depth of device – 5 m, distance device – hydrophone – 10 m

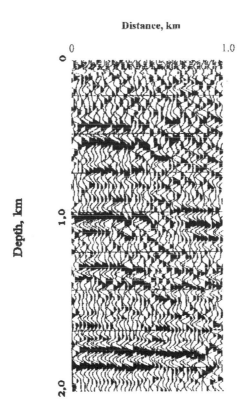

Figure 6. Depth cross-section form Perm region

PHYSICAL AND CHEMICAL PRINCIPLES OF SLURRIES

CARL-OTTO LEIBER *, RUTH M. DOHERTY **
c/o Fraunhofer-Institut für Chemische Technologie (ICT), Joseph-von-Fraunhofer-Str. 7, D 76327 Pfinztal (Berghausen), Germany – formerly BICT/WIWEB, Swisttal-Heimerzheim* Naval Surface Warfare Center, Indian Head, Maryland, USA.**

1. Introduction

There are many names for an old and inexpensive type of explosive that is now finding modern applications. It has mostly been used in bulk for commercial applications, but has also been considered for some military applications. We refer here to what are called Slurry Explosives. The value of employing a slurry was discovered when it was observed that some water-soluble explosives, when in contact with incidental water in a bore hole, exhibited surprising power compared to the original explosive, ANFO (**A**mmonium **N**itrate/**F**uel **O**il). The science of Slurries is intimately linked with the name of MELVIN A. COOK, who edited a book [1] on this subject. As slurries were developed into a sophisticated product, the name of the product also changed, so that now names such as Blasting Agents, Water-Gel Explosives, Emulsion Explosives, and so on may be used to refer to such materials.

Historically commercial explosives have been linked mainly with nitroglycerin, which was first made by SOBRERO in 1847, and manufactured by NOBEL in 1852. However, as early as 1867, the combination of an oxidizer with a fuel was patented to OHLSON and NORRBIN in Sweden. In 1885 a patent for a fuel-oxidizer composition was issued to PENNIMAN (for ANFO) in the US. In 1891 DE BRUYN reported the surprising finding that pure AN could be detonated. This finding was later supported by LHEURE (1907), ROBERTSON (1914), and the US-BUREAU OF MINES [2]. In Russia in the 1930s GRIGORY DEMIDYUK investigated such explosives [3]. But it was the Texas City accident in 1947 that brought the power of Ammonium Nitrate for commercial explosives to widespread attention.

2. On Explosives

This historically important incident teaches us that an explosive need not be related to any unstable or metastable molecules, nor to any super-high energy content. It is the liberation of the available energy within a short time that is essential. An explosive is, therefore, an energetic substance with the internal ability of power conversion. This is demonstrated by the following example. One kg of nitroglycerin (NG) possesses the energy content of about 100 g of oil burning in air. A 1-kg sphere of NG (diameter 10.63 cm) is consumed (theoretically) by burning within about 10 minutes, whereas in a detonation, the reaction is complete within about 7 µs. In the case of detonation we have, therefore, a power conversion by a factor of the order of 600 s/7 µs ≈ 86 million! In electronics there is not a single amplifier that can provide such a magnification factor!

This picture readily highlights the main points of concern for any explosive:

1. Maximizing the chemical energy density, in combination with
2. Controlling of rate of energy liberation or power conversion.

The main goal of any scientific consideration is not so much finding an applicable technical solution as it is understanding the basic scientific aspects that can potentially be used to tailor the properties of any given system. In laboratory slang, we might refer to this as identifying the knobs that can be used for tuning the explosive output.

3. Energy Sources

Most powerful practical energy-liberating systems are based on combustion reactions, usually involving aluminum. However, aluminum produces mostly condensed reaction products in the final state, whereas the combustion of carbon and hydrogen is favorable for both the heat of combustion and the volume of gaseous products. Nitrogen is not as favorable, since this behaves almost as an inert or leads to toxic products. For environmental reasons, as well as thermal reasons, halogens are not ideal components as combustibles or as oxidizer.

As the oxidizing species, oxygen and fluorine are favorable, but when cost, availability, and environmental aspects are considered, only oxygen is practical.

In order to get a high energy density, the density of the fuel and the oxidizer should be as high as possible.

NG as a 'classical explosive' contains in its molecular structure carbon, hydrogen, nitrogen, and oxygen so that the oxygen within the molecule can react with the carbon and hydrogen in the molecule to form the final products. Since the oxygen is available in the same molecule as the oxidizable species (C and H), no diffusion processes are necessary to bring together the components to react. If, however, the fuel and the oxidizer are separated from one another, transport properties govern the rate of the combustion reaction. A qualitative description of this situation is given by the word *intimacy*. Obviously this intimacy is greatest for a molecular explosive, like NG. This is followed by an intimately mixed medium in which the oxidizing species is mixed with a fuel (intermolecular explosive). At the far end of the spectrum, one finds the bulk oxidizer and bulk fuel separated in two containers. The last of these is of great practical importance, since, in principle, any explosive can be transported safely in two components insofar as the components cannot detonate by themselves.

3.1 SAFETY PROBLEMS OF BINARY EXPLOSIVES

In binary explosives, the energy producing reactions are mainly or exclusively between two separate species. The fuel components can be completely oxygen-free or may contain some oxygen, but at a low level (extremely negative oxygen balance). They are not detonable and are assigned to the UN-classes 3 (flammable liquids) or 4 (flammable solids), but not to UN-class 1 (explosives).

Although in the past there have been cases of liquid oxygen (or liquid air) being used as the source of oxygen in some explosives, the high sensitivity and unpredictable results made this impractical, and such formulations are no longer used. As an alternative, other oxygen-containing compounds are used as oxidizers. This oxidizer is an oxygen-rich compound characterized by a more or less high oxygen balance, often attached to a nitrogen or chlorine atom by a relatively weak bond (in contrast to oxygen in a stable, low-energy compound such as alumina (Al_2O_3), for example). The preferred oxidizers of practical interest are ammonium nitrate and ammonium perchlorate. There are many nitrates, chlorates, perchlorates, chlorites and nitrocompounds, like tetranitromethane, that could be used. None of these belong to the UN-class 1. Nevertheless, in special testing outside scope of the UN-defined procedures, most of them reveal some detonation hazards, at a minimum the risk of Low Velocity Detonation.

3.2 ENVIRONMENTAL RISKS

Ammonium perchlorate (AP) is a useful oxidizer, but has the disadvantage of containing chlorine, and thus producing chlorine-containing compounds, such as hydrogen chloride or dioxins, in the product gases. At least in the western world, there is such a strong political aversion to even the word "dioxin", that it is recommended that serious consideration be given to the political implications if the use of AP is envisaged. Furthermore, all chlorates and perchlorates poison the soil. Due to the possibility of slurries failing to detonate in a bore hole, the risk of soil contamination is an additional reason for concern. In summary, it can be stated that AP-based explosives may be highly energetic, but their use can introduce significant problems for environmental reasons.

3.3 MIXED SALT SOLUTIONS

In commercial applications, it is common to use slurry explosives that contain a mixture of oxidizing salts in solution. Among those compositions in use are solutions based on combinations of AN and calcium nitrate (CN), and of AN and sodium nitrate (SN) [4]. The advantages of these mixtures are found in the flexibility that they allow in tailoring the energy content, detonability, and critical diameter of the explosive slurry. Some other combinations of oxidizer salts, such as a mixed nitrate/perchlorate salt solution (e.g., AP and CN) could also yield superior results to the use of a solution of a single oxidizing salt, owing to the weak dependence of the solution of one salt on that of another salt with which it has no ions in common.

3.3.1 *Risk of such procedures*
Care must be taken to avoid compatibility problems if mixed salts are used to increase the available oxygen. There is a known incompatibility between the ammonium cation and the chlorate anion, although this may be mitigated by the addition of a phosphate(anion) in water [5]. This incompatibility does not apply only to the deliberate mixing of ammonium salts and chlorate salts, since several 'spontaneous' explosions involving the storage of commercial explosives based on AN and sodium nitrate with sodium chlorate impurities (not in solution) in the crystalline state [6] have occurred. BRETHERICK [7] states that ammonium chlorate always explodes spontaneously above 100°C.

3.4 COMBUSTIBLES

The fuel in a fuel-oxidizer combination can be nearly any oxidizable material, with carbon- and hydrogen-rich components being preferred. Thus any oils, and even wood-meal, can be used, as well as waste organic materials such as demilitarized high explosives and propellants. Aluminum as a fuel has the advantage of a high energy of combustion. AP-based rocket propellants are less favorable, due to the presence of the AP (see above).

3.5 MAINTAINING THE MIXTURE OF FUEL AND OXIDIZER

Water, which can be used as a solvent, behaves as an inert, and is therefore only a medium for delivering the other ingredients. It is therefore desirable to reduce the water content as much as possible in order to keep the energy content of the formulation high. To maintain a controlled „intimacy" of mixture between the fuel and the oxidizer, slurries are designed to have a high viscosity which may be achieved by the use of a thickening agent or stabilizer. This can also be achieved by the **emulsion** technology, where, by the use of emulsifying agents, the original conditions of dispersal are maintained. The continuous phase usually is the fuel (water-in-oil emulsions).

Another option is to use a eutectic melt of, for example, ethylenediamine dinitrate (EDDN) with ammonium nitrate (AN), where a eutectic melt is obtained for EDDN/AN = 49/51 by weight, melting at about 103°C. The advantage of this type of explosive is that the formation of the eutectic places the fuel-rich salt (EDDN) and the oxidizer-rich salt immediately adjacent to one another. Such types of explosives, and also EAK-explosives based on of EDDN/AN/KNO$_3$ were even evaluated for military applications [AKST [8, 9], GUPTA [10]]. For small charges their insensitivity was favorable, but their blast performance was not, whereas for large charges Comp B-like performance was achieved, but in very large scale shock sensitivity testing, they were very easy to initiate (low pressure), provided the charges were long enough to allow for growth of reaction. The conclusions about sensitivity that were drawn from small-scale tests were misleading, therefore.

Emulsion techniques have also been applied in the solid state for the control of the intimacy, where the fuel content was of the order of 10%, emulsified in the melt. Due to super-cooling proper emulsions were also very homogeneous at room temperature. Such castable emulsions were free of water, and it was possible to machine them.

4. Power conversion

A relative compression $\Delta v/v_0$ can be realized by both static and dynamic means. In plane wave terms this compression is given by the ratio of the particle velocity and the shock velocity. According to STEVERDING [11] the particle velocity corresponds to a group velocity and the shock velocity to a phase velocity. These velocities are related by the relations of dispersion, and by the REYNOLDS-RAYLEIGH-relations of hydrodynamic energy transport [12, 13] (see also the discussion of SOMMERFELD [14]). LEIBER [15] used this approach to illustrate the basic differences between dynamic and static high pressure effects.

$$\frac{\Delta v}{v_0} = \frac{u_p}{u_s} = \frac{\text{group velocity}}{\text{phase velocity}} = \frac{\text{Energy flux per unit area in time } \tau}{\text{Energy content ahead in unit area and length of } u_s \tau}$$

Detonation is a hydrodynamic power conversion. Due to the relations of dispersion we must take into account hydrodynamic phenomena, which correspond to an anomalous dispersion. This addresses the question of detonation precursors, which necessarily require a two-phase system, where the phases are characterized by significantly different densities and compressibilities.

In the case of a normal dispersion only part of the generated energy is propagated in the direction of the wave, whereas in the anomalous case the system is exhausted due to the tendency to transfer more energy than is generated. In other words, the anomalous dispersion is a natural brake to prevent achieving "high dynamics". This is a formulation in hydrodynamic terms of the detonic phenomena of Low Velocity Detonation (LVD) and High Velocity Detonation (HVD), with distinct different mechanisms and efficiencies.

4.1 DIFFERENT COMPRESSIBILITIES (BUBBLES)

The importance of heterogeneities for the detonability of a wide range of explosives from TNT to high NG-content commercial explosives is very well known. It is manifested in the phenomenon of deadpressing, whereby the strength of the stimulus required to initiate an explosive increases as the void content decreases, and correspondingly the % of theoretical maximum density approaches 100%. At some point the explosive cannot be initiated – it has been deadpressed. This effect is tremendously important for slurries. Large inhomogeneities in density and compressibility can be achieved in slurries through the introduction of bubbles and voids by a wide variety of mechanisms. Among the modern tools that can be used to incorporate voids or bubbles are such methods as aeration, chemical gassing, and the addition of polystyrene balls or glass microballoons. The technique of introducing bubbles or inhomogeneities into commercial explosives and even blasting gelatin by squeezing or kneading is known to make charges that are too homogeneous or have become insensitive during storage detonable. The void percentage typically used in practical applications may reach 20 Vol%. Another much more sophisticated mechanism, that of the hydrodynamic pair formation, will be addressed below.

Such voids are collapsed rapidly by the priming charge, leading to a gigantic power conversion quite similar to a Rayleigh collapse. This collapse leads to the emission of pressure and tension waves from the single voids (see Figure 1).

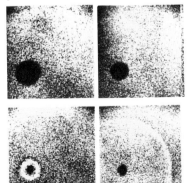

Figure 1:

Collapse of a LASER-induced cavity in a sequence of time distance of 1 µs. The size of a picture corresponds with 2.7 mm. The radiation of pressure (tension) waves is clearly seen.

(Courtesy of LAUTERBORN and TIMM [16])

4.2 BUBBLE ARRAYS AND THE STRUCTURE OF DETONATION FRONT

Since the process of bubble collapse and expansion is time-dependent, we use in the following discussion harmonic motions to demonstrate the effects of various bubble arrays. Fig. 2 shows the pressure-distance distribution of one finite source, where all geometric quantities are related to the harmonic wavelength λ, so that the length scales are dimensionless, and represent geometric/dynamic relationships.

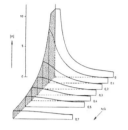

Figure 2:

Pressure/dimensionless distance profile of *one* harmonic finite source (bubble).

Considering now an array of harmonic pressure sources, one obtains macroscopic pressure profiles. Figure 3 demonstrates the variability of the pressure profiles of the same sources from smooth to rough, but in different geometry and numbers. Figure 4 shows such a rough structure of the detonation front of NM/acetone mixture (35/65 by weight). Finally Figure 5, top, shows one detonation front, the left part smooth, and the right part rough, where in the dividing part a dark wave is present. A detail of the top is shown in the center, which is modeled by specific arrangements of the sources in the bottom. These Figures indicate that all macroscopic detonation structures can be constructed by proper source arrangements.

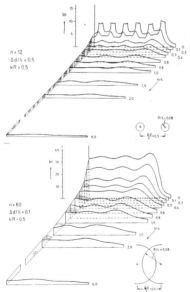

Figure 3:

Variability of the pressure/distance profiles of an array of sources (Figure 2) depending on the dimensions and number of sources.

Above: Wavy profile, where the character of the source is manifested. 12 sources are in a linear array, and their size and distance are shown on the right hand edge.

Bottom: Array of 60 synchronous sources in the same diameter as above. Since the expanding bubbles penetrate each other (foaming up) a smooth but wavy profile results, where the character of the source is completely hidden. This situation is shown on the right hand side.

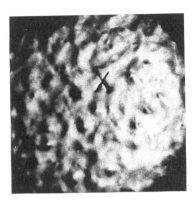

Figure 4. Rough structured detonation front of a just detonable NM/acetone mixture 35/65 by weight as observed by the impedance mirror technique developed by Dean Mallory[18]. The X corresponds to a height of 3.2 mm. Note, that the detonation front of pure NM is smooth and practically plane.

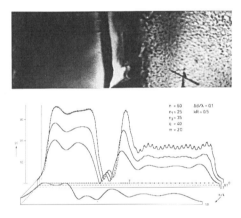

Figure 5. Modeling of the transition from a smooth to a structured front, and in the center a dark wave. Such phenomena were observed by the impedance mirror technique of H. Dean Mallory[19] for a NM/acetone mixture 80/20 by weight. The smooth/rough front is modeled with varying source distances, and the dark wave results from the absence of the sources.

4.3 CELL AND HERRINGBONE STRUCTURE OF DETONATION

The detonation in gases shows cell structures, whereas condensed explosives show herringbone structures at the sides, see Fig. 6.

Figure 6. Herringbone detonation structures of detonating nitromethane in a plastic tube. The scale is in cm. (bottom)

The assumption of harmonic sources is too simple to allow modeling of such behavior. . More realistic is the following behavior of a bubble: Assume a bubble collapsing (or expanding) for a short period of time, after which its motion stops. The radiated pressure *and* tension waves become distance *and* time dependent, see Figure 7.

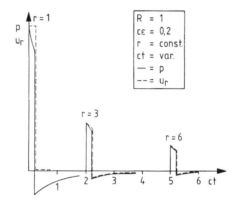

Figure 7. Radiation of pressure and tension waves by a single bubble, that expands for only a short time. Pressures and radial velocities are shown for a constant distance r and a varying time t. For more details, see [20].

Bringing this source of Figure 7 into an areal array of 13 sources, see Figure 8, then one gets both cell- and herringbone-structures. Since the amplitudes of the very regular cells are small, these are visible in low amplitude pressures of gas detonation, but not in high amplitude detonation pressures of condensed explosives. In the latter case these cells disappear as by a superpressurized stampmark. Contrary to the cells the herringbone structures are deep, and do not disappear. We have modeled structures of detonation without any assumptions simply by appropriate arrangements of the appropriate sources [20].

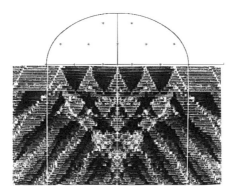

Figure 8. Modeling of heringbone detonation structures in a 2-dimensional array of 13 sources as shown in Figure 7.

4.4 ONSET OF THE CHEMICAL REACTION

By compression (or expansion) of the bubbles dominant shear losses on the surface result, as well as adiabatic heating (or cooling) of the bubble contents. The latter mechanism is known as Bowden's hot spots, which, however, can be negligible compared with the viscous loss power still in the medium without any heat transfer. The loss power amounts to many MW/cm³, and therefore drives any chemical reactions in the surrounding medium if at all possible! Any Arrhenius kinetics become unimportant. Figure 9 shows an igniting sequence in NG from COLEY and FIELD [21, 22].

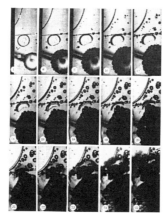

Figure 9. Bubble in a nitroglycerin (NG)-film in 1 μs-intervals. The onset of chemical reaction is clearly seen at the bubble surface. E is the electrode for electrical ignition. W is the thin wire for holding the bubble, and F is the boudary of the NG-film. (Courtesy of G.D. Coley and J.E. Field [21, 22]).

4.5 DIFFERENT DENSITIES (BUBBLES AND PARTICLES)

Assume a particle or void of density ρ' in a medium of density ρ_∞, which is pushed by a shock pressure. Then there are 3 cases:

- **Case $\rho' = \rho_\infty$**: In this case there is homogeneity, provided that the compressibility is also the same.

- **Case $\rho' < \rho_\infty$**:

  Due to the conservation of momentum the particle velocity of the less dense particle becomes larger than that of the surrounding medium. A relative particle velocity between the particle and the medium results. Even a void obtains asymptotically 3 times the particle velocity of the medium. Figure 10 gives an example of this, where a void in a shocked high density, high strength material shows traces like those from a shot. It should be noted, that such holes completely disappear with hot forging!

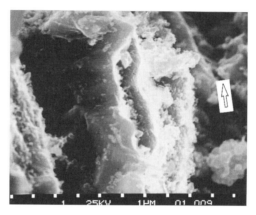

Figure 10. Shock activated bubble mobility in molybdenum. The shock of the order 375 kbar in the direction of the arrow induced this mobility, whereas such bubbles completely disappear by hot forging.

4.5.1 Void Precursors, Onset of High Velocity Detonation

With bubble velocities $u'_p = 3\, u_{p\infty}$ and the given Hugoniots, there exist dynamic ranges wherein the bubble velocities overtake the material shock velocities, and precursors result. If such precursors penetrate the shock front they expand ahead of the pressure front and create both luminosity and new pressure centers ahead of the main pressure front. Experimental examples are shown in Figure 11 and 12.

Figure 11. Detonating ANFO in water (Diameter 50 mm, D = 2,300 m/s). Note the shock front in water, and the luminosity ahead the pressure front (Courtesy of Fossé [23]).

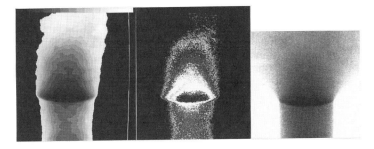

Figure 12. Detonation head of a foamed PETN/PU = 60/40, ρ = 0.6 g/cm³. Left: X-ray-flashed detonating charge. From the left picture photographic equidensities were produced, which clearly demonstrate that pressure precursors exist ahead of the detonation front. Due to the cylindrical geometry of the charge these precurors are visible only in the center of the charge. Right: The same pictures with digitized densities [24,26].

All these mobilities depend strongly on the particle sizes in comparison with the shock rise length (dynamics)! The onset of these precursors is the criterion of the initiation of High Velocity Detonation. For liquid explosives the predicted onsets compare favorably with experiments, even for hydrazoic acid [25, 26].

This precursor mechanism shows also an inherent synchronization of bubble sizes and phases, so that in contrast to LVD-phenomena the HVD-phenomena are quite regular [25].

- **Case ρ' > ρ∞:**

Due to the conservation of momentum, the particle velocity of the heavy particles lags behind the particle velocity of the matrix. Figure 13 shows an experiment where tungsten particles in aluminum impact upstream only of the reference area. But there appears an additional and most important effect:

4.5.2 Hydrodynamic Pair formation
As the dense particle lags behind the matrix material, there arises a relative velocity of the particle with respect to the matrix. This particle flow can be laminar (material closes again after the denser particle) or turbulent. In some cases there can be wake formation,

depending on the Reynolds number of the flow. In the latter case, due to the particle displacement by a shock, a void is formed. This newly generated void (in an originally voidless medium) can be driven into the direction of shock, as shown in Figure 10. Such a situation is shown in Figure 13.

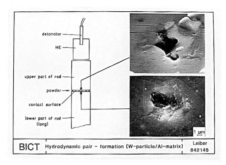

Figure 13. Model experiment for the hydrodynamic pair information. There is an aluminium rod (ρ^∞ = 2.7 g/cm3), where between the upper and lower part (with polished surfaces) tungsten (ρ ' = 19.3 g/cm^3) particles of about 1 μm size are inserted. By the shock attack these produce impact holes in the upper part, whereas in the lower part cavitation damages occur.

This hydrodynamic pair formation is of the utmost importance, since bubble sensitization becomes possible in a fully dense material having only dense inerts. In comparison to sensitization by deliberate inclusion of voids, where the voids can dissappear with increased static pressure, this hydrodynamic pair formation works under any high pressure conditions, so that even seismic explosives can be sensitized.

4.5.3 Problems of dynamic particle agglomerations

If there are many particles with different relative velocities there is a tendency for them to agglomerate and move together due to the Bjerknes forces (Bernoulli effects). An example of such a behavior is given in Figure 14. In the case of slurries a reduced chemical reactivity results.

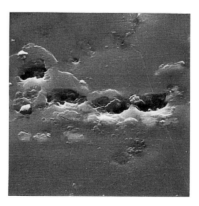

Figure 14. If the tungsten particles are too closely spaced (agglomerations) between the surfaces, then coalescence appeared. The postulated laminar and turbulent flow is also seen in the lower part of the picture, where the holes closed again.

4.5.4 Ahrens Selectivity

As an essential leading idea of the science of mixed intermolecular explosives AHRENS introduced the principle of selectivity. This idea is described best by the particle mobilities in a dynamic activated mixed explosive. Depending on the viscosity of the matrix, which tends to fix particles in place, and the dynamics, particles of different densities move ahead or lag behind the main matrix average. As a result, a dynamic demixing may occur. This can be used as a tool to tailor reactivity, to determine whether constituents react in the detonation zone or at a later time. In a well-confined charge the net result is a time average of the reaction, whereas in the case of the rupturing of the confinement unreacted particles can be released into the surrounding medium, where they can react with the surrounding air or behave as inerts. One example is aluminum. Very fine aluminum can react in the detonation zone, whereas coarse aluminum may simply be heated up, to react later with detonation product gases, air or water. In safety in mines explosives there is another pairing essential: Ammonium chloride and sodium- or potassium nitrate. Within the detonation zone or an intact bore hole a reaction may occur between the nitrate ion in the alkali metal nitrate and the ammonium ion in ammonium chloride, which is equivalent to the decomposition reaction of AN. However, in a freely suspended cartridge in air the salts are dispersed with little or no reaction into the surrounding air (as inerts). Therefore, depending on conditions, the liberated energy can be varied greatly! The original idea of AHRENS is given in [27], and that based on dynamic particle effects in [28].

Examples of grain size dependent mobilities in a matrix are given by the mobilities of carbon in cast iron. By shocking cast iron carbon is transferred to an attached metal plate. Another situation is, that in the case of shocked cast iron the carbon particles move like a wedge through the matrix resulting in the complete destruction of the sample [29]. Figure 15 shows a summary of these effects.

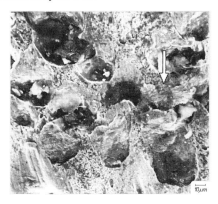

Figure 15. Shocked cast iron with spheroidal graphite particles tending to remain on the spot (upper left) to acquire a mobility into the shock direction, indicated by the arrow (small particles), upper left, and a reverse jetting in the center. From the particles sizes remaining on the spot, the shock rise lenght can be concluded.

4.5.5 *Selectivity by Particle Coatings*

Additional selectivity by surface coating is also feasible. Some examples are given in [30]. Reactive or even inert chemicals of about 0.03% by weight coated on the surface of particles can greatly influence the power of commercial explosives by physical scattering effects. If the same amount of the chemicals is added uniformly to the matrix practically no effect results, as one would also expect from current code calculations.

4.5.6 *Critical Dimension Phenomena*

In every case, a single hot spot radiates its pressure isotropically. But in arrangements of such sources according to dynamics, directional patterns that depend strongly on the source configurations appear [31]. As a result, the critical diameter in a cylindrical geometry has no generally applicable characteristics. The critical dimensions of thin sheets or surface coated (empty) tubes are much different from those of the classical cylindrical critical diameter.

Therefore the critical diameter in cylindrical geometry is not the only measure that is relevant to safety.

For example, in the course of investigations of nitromethane (NM) about 200 experiments were performed in conical geometry. The conventional expectation is, that at the critical diameter (15 mm at room temperature) the HVD dies away. But in experiments this was true only in about 80% of all cases. In nearly 20% of the cases at the critical diameter a transition from detonation to burning occurred. In 2 cases, however, at the critical diameter (in plastic confinement!) a transition from HVD to LVD occurred, which again transited to HVD, and then to LVD, which finally died away at a diameter of 2 to 3 mm. This behavior was observed by electric probes, by photographic records, and by plating experiments, all simultaneously [26], see Fig. 16.

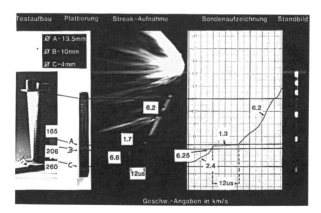

Figure 16. "Pathological" events at the critical diameter of NM in plastic confinement, where transitions HVD ⇨ LVD ⇨ HVD ⇨ LVD ⇨ fading out was observed.

5. Problems of Slurries

5.1 STRUCTURAL STABILITY OF THE EXPLOSIVE

It can be seen from the description above that the structural stability of the explosive is of the utmost importance. Coagulation of particles or settling of solids in the slurry can lead to serious deviations from expected behavior. Therefore the viscosity of the matrix is of great importance. It can be tailored to some extent by crosslinking agents. Due to the very limited time of storage (some slurries are used immediately upon formulation), and the proprietary nature of commercial explosives, there is ordinarily a recommended range of temperatures within which the explosives will maintain their structural integrity.

5.2 TEMPERATURE DEPENDENCE OF THE PROPERTIES

In addition to the dependence of the chemical properties on temperature, there is also a strong dependence of the structural properties of slurries on temperature. Therefore with varying temperatures the detonation properties may also vary.

5.3 CHEMICAL REACTION IN DETONATION AND POST DETONATION REACTIONS

Due to the dynamic particle effects, which strongly depend on the particle sizes, the reaction in detonation can be greatly influenced, intentionally or unintentionally. The reaction can result in the detonation zone, or in the product gases, which can in addition use the ambient air (or water) as additional oxidizer. In such cases the power of brisance can greatly vary in combination with a corresponding variation of the blast.

5.4 DEPENDENCE OF DETONATION ON STATIC SURROUNDING PRESSURES

The sensitization of slurries depends on the presence of a sufficiently high concentration of voids. The sensitivity of a slurry explosive to ambient static pressure depends on the means of void generation. In the case of chemical gassing, an increase of static pressure not only compresses the voids, but also makes them vanish by dissolution in the matrix. Therefore ordinarily a slurry-type is usable only once. That is to say, if a bore hole is filled and for some reason is emptied again, refilling the bore hole with the same explosive leads to duds due to loss of porosity. This is also the reason why the use of microballoons may be preferable. A more detailed description of the pressure dependence of initiation and detonation is given by LEIBER [32].

5.5 INTERACTIONS BETWEEN CHARGES (GERMAN: SCHUßBEEINFLUSSUNG)

Usually the charges of slurries fill up the borehole without any spaces, so that channel effects [33] can be ignored. But if two bore holes are too close to one another, it is possible for one detonating charge to compress a second charge by one or more of several possible mechanisms. The results can be manifested in many different ways: deadpressing (i.e., no initiation), premature detonation, or even deflagration or burning. This effect is of concern mainly if there is a time delay in initiation.

5.6 CRITICAL DIAMETER PROBLEMS

In contrast to military explosives, where the critical diameter below which detonation fails, is ordinarily of the order of several mm, the critical diameter in the case of slurries may be of the order of cm or dm. The critical diameter also depends on the confinement. It is therefore indispensable, that this critical diameter is smaller than the diameter of the borehole. In practical terms, this means that the slurry in use must be suitable for the bore hole diameter.

5.7 PROBLEMS OF SENSITIVITY TESTS AND ASSOCIATED SAFETY PROBLEMS – REGULATIONS [34]

The reason for performing sensitivity testing should be to enable one to predict hazards. But even with the insight available to us today, the famous disasters involving AN-explosions (Oppau, 1921, Texas City, 1947, and others) could not be foreseen by classical conventional explosives sensitivity testing alone. The same applies to non-UN-class 1 substances, such as pure chlorates, hypochlorites, nitrates, etc. Since, in current testing, ANFOs and slurries show no response that is identified with explosives, these would not be assigned as explosives – if it were not for the fact that they are used as typical explosives. Therefore they are considered as 'insensitive' explosives. A special UN-class was created for these materials, for transportation purposes: UN-class 1.5 for the substances themselves, and UN-class 1.6 for articles (ammunition) containing these materials. Specific tests are defined for these classes. It is stipulated that the probability of any accidental explosion (and associated cratering) is negligible, so reduced safety distances, based on the expectation of mass fire rather than mass detonation, are thought to be adequate.

There are practical benefits to the use of UN-class 1.5. Transportation is less expensive, as are insurance fees, and storage and mass handling with reduced caveats are also much cheaper. In light of the economic benefits associated with this class, serious consideration should be given therefore to whether it is appropriate to treat these materials differently from other classes of explosives. For the following reasons we believe that it is not acceptable.

5.7.1 *Accident Histories*
The comparison of explosion hazards shows that there are a greater number of explosion hazards of 1.5-substances compared to explosives such as TNT [34]. (To be fair, it should be noted that much larger quantities of commercial explosives are produced than classical 1.1-explosives.) Contrary to the expectations of the regulations even ANFO cratered in the accident of Kansas City (1988).

Other explosion hazards have been observed around the world with **non**-UN-class 1 substances. Some of these have occurred spontaneously (by marshaling – i.e., handling in railway yards), but usually they are caused by external fires. Examples are nitromethane explosions, 1958 (UN-class 3), MMAN-explosions (marshaling), and explosions of pure chlorates, perchlorates and even bleaching powders by external fires [34].

5.7.2 Explosion Accident Involving a Slurry - Herlong, CA, USA, 07.08.1978 [35]

This incident involved an assembly of 10 bombs filled with 12,600 lb. IRECO-Slurry DBA-22M. The slurry composition was a mixture of AN/Al 50/34, with the rest being water with guar gum, boric acid, and ethylene glycol. The assembly had been stored for a long time (1972 to 1978) in open air. The bombs were 1.35 m in diameter and 3.20 m long. After a spell of very hot weather (four days in a row around 40°C), detonation occurred on the morning of the fifth day at 8:38 am. Various witnesses described double explosions, and others mentioned sequential events with time intervals of about 1 to 2 seconds: first, a flash of fire with a pop or pops, or loud popping noise; second, a mushroom-shaped cloud; and third, a disastrous explosion. Many rationales for the occurrence of the event were discussed, but no definitive conclusion was reached.

6. In Summary

Slurry behavior is an example of applied hydrodynamic two-phase principles, where the knobs to be adjusted to tailor performance are the densities and compressibilities of the components, as well as the dynamics. Particle sizes and the fixing forces of the matrix are essential factors in achieving the desired performance. Due to the enormous hydrodynamic loss power, classical Arrhenius kinetics is not relevant for the onset of chemical reaction. Each composition can be forced to react.

Such two phase systems exhibit the phenomena of critical diameter and critical paths of detonability, such as moderate static pressures. Ambient pressure or very moderate static pressures of the order of one bar, can lead to the disappearance of voids. Therefore we have in the slurries both very robust and very sensitive principles at work that can be varied over a wide range to meet the needs of a specific application.

7. References

1. Cook, M. A. (1974) *The Science of Industrial Explosives*, IRECO Chemicals, Salt Lake City, Utah.
2. Munroe, Ch. E. (1923) Die Explosivität des Ammonnitrats, *Zeitschrift für das gesamte Schieß- und Sprengstoffwesen* **18**, 61/64.
3. Kondrikov, Boris, Email communication from 10.09.99.
4. Held, Martin, Westspreng GmbH, FAX-communication from 22.05.00.
5. Hirosaki, Y. and Hattori, K. (1984) Stabiles wäßriges Oxidationsmittel, *German Patent DE 3243926 C2*.
6. Schultze-Rhonhof (1944) Gutachtliche Äußerung über die Sprengstofflagerbrände auf der Grube Eisenzecher Zug in Eiserfeld bei Siegen am 7. und 9. Februar 1944 sowie die Sprengstofflagerexplosionen auf der Grube Sachtleben in Meggen am 9.2.1944 und auf der Schachtanlage Fröhliche Morgensonne in Wattenscheid am 12.2.1944. *Berggewerkschaftliche Versuchsstrecke, Dortmund-Derne, Tgb.-Nr. 705/44 SR/R*.
7. Bretherick, L. (1979) *Handbook of Reactive Chemical Hazards*, 2nd ed., reprint 1981, Butterworth, London, Boston, Sydney Wellington Durban Toronto, ISBN 0-408-70927-8.
8. Akst, I. B. (1981) Detonation in Intermolecular Explosives: Characteristics of some Eutectic Formulations, *7th Symposium (International) on Detonation, Annapolis, Maryland, 548/559. NSWC MP 82-334*.

9. Akst, I. B. (1985) Intermolecular Explosives, *8th Symposium (International) on Detonation, Albuquerque, NM, 1001/1010, NSWC MP 86-194*.

10. Patrick, M., Gupta, A. and Harden, S. (1991) Intermolecular Emulsion Explosives, *US-Patent # 4994123*, Intermolecular Complexes, *US-Patent # 4948438*, Intermolecular Emulsions, *US-Patent # 5145535*.

11. Steverding, B. (1972) Quantization of Stress Waves and Fracture, *Mater. Sci. Eng.* **9**, 185/189.

12. Reynolds, O. (1877) *Nature* **46**, 378; *Collected Works, Vol. I, 198*. - Reference from 13) and 14).

13. Lord Rayleigh (1877) On Progressive Waves, *Scientific Papers, Vol. I, # 47, 322/327*. Dover Pub., New York 1964.

14. Sommerfeld, A. (1945) *Mechanik der deformierbaren Medien*, pp. 179/185, Akademische Verlagsgesellschaft Becker & Erler, Leipzig.

15. Leiber, C. O. (1977) Basic Differences between Dynamic and Static High-Pressure Effects, *High Temperatures – High Pressures* **9**, 573/574.

16. Lauterborn W. and Timm R. (1980) Bubble Collapse Studies at a Million Frames per Second, in Lauterborn, W. (ed.), *Cavitation and Inhomogeneities in Underwater Acoustics*, Springer, Berlin. Heidelberg, New York, pp. 42/46.

18. Mallory, H. D. (1967) Turbulent Effects in Detonation Flow: Diluted Nitromethane, *J. Appl. Physics* **38**, pp.5302/06.

19. Mallory, H. D. and Greene, G. A. (1969) Luminosity and Pressure Aberrations in Detonating Nitromethane Solutions, *J. Appl. Physics* **40**, pp.4933/38.

20. Leiber, C. O. (1993) Detonation Model with Spherical Sources J: Cell- and fish bone Structures in Detonation. *Europyro 93 – 5e Congrès International de Pyrotechnie du Groupe de Travail, Strasbourg, France*, pp. 497/509.

21. Coley, G. D. and Field, J. D. (1973) The Role of Cavities in the Initiation and Growth of Explosion in Liquids, *Proc. Roy. Soc. London* **A 335**, pp. 67/86.

22. Coley, G. D. and Field, J. D. (1973) The Sensitisation of Thin Films of Nitroglycerin, *Combustion and Flame* **21**, pp. 335/342.

23. Fossé, C. (1966) Influence de l'amorçage sur la pression exercee dans l'eau par les explosifs industriels, *Comm. XXXVIe Congrès International de Chimie Industrielle, Bruxelles*, pp. 28/32.

24. Hessenmüller, R. and Leiber, C. O. (1984) Wie eine Detonation entsteht, *Die Umschau* **84**, pp. 507/509.

25. Leiber, C. O. (1991) Detonation Model with Spherical Sources F: Dynamic Void Mobilities, Alterations of the Hugoniots by Bubble Flow, *Proceedings of the 17th International Pyrotechnics Seminar combined with the 2nd Beijing International Symposium on Pyrotechnics and Explosives, Vol. I*, pp. 733/757, Beijing Institute of Technology Press.

26. Leiber, C. O. (1991) Detonation Model with Spherical Sources G: Dynamic Void Mobilities, HVD Initiation of liquid Explosives, *Proceedings of the 17th International Pyrotechnics Seminar combined with the 2nd Beijing International Symposium on Pyrotechnics and Explosives, Vol. I*, pp. 722/732, Beijing Institute of Technology Press.

27. Ahrens, H. (1956) Die Bedeutung des selektiven Verhaltens der Detonationswelle im Gebiet der Wettersprengstoffe, *Explosivstoffe* **4**, pp. 102/109.

28. Leiber, C. O. (1979) Zur Ahrens' Selektivität, Ansatz zu einer Theorie gewerblicher Sprengstoffe, *Nobel Hefte* **45**, pp. 65/76, *Nobel-Hefte* **45**, p. 167.

29. Leiber, C. O. (1975) Dynamic Particle Motion in Materials as a Consequence of the finite Shock Rise, *5th Int. Conf. on High Energy Rate Fabrication, Denver, CO, USA*, pp. 1.5.1/1.5.10.

30. Leiber, C. O. (1997) Detonation Model with Spherical Sources M: On the so called *non-ideal* Behavior of

Explosives, . *Proc. 23rd Int. Pyrotech. Seminar, Tsukuba, Japan,*.pp. 441/456.

31. Leiber, C. O. (1994) Detonation Model with Spherical Sources K: Critical Dimension Phenomena, - Asymptotic Considerations, *Proc. 19th Int. Pyrotech. Seminar, Christchurch, New Zealand*, pp. 40/51. (CA 122:59576).

32. Leiber, C. O. (1986) Approximative quantitative Aspects of a hot spot, Part II: Initiation, Factors of safe handling, reliability and effects of hydrostatic pressure on initiation, *J. Haz. Mat.* **13** pp. 311/328. CA: 105(22):193919z

33. Leiber, C. O. (1990) Detonation Phenomena in Charges with an axial Hole, *Annex II, Hazard Studies for Solid Propellant Rocket Motors (Etudes des Risque pour les Moteurs-Fusées à Propergoles Solides), AGARDograph No. 316, AGARD-AG-316*, pp. 167/168.

34. Leiber, C. O. and Doherty, R. M. (1998) Review on Explosion Events - a Comparison of Military and Civil Experiences – in V. E. Zarko, V. Weiser, N. Eisenreich and A. A. Vasil'ev (eds.): Prevention of Hazardous Fires and Explosions, The Transfer to Civil Applications of Military Experiences, NATO Science Series, 1. Disarmament Technologies Vol. 26, KLUWER Academic Pub, Dordrecht, Boston, London, 1999, ISBN 0-7923-5768-X (HB), ISBN 0-7923-5769-8 (PB), pp. 1/15.

35. (1978) Report of Proceedings by Investigation Board of the BLU-82/B Bomb Explosion and Explosion of BLU-82/B Warheads and related EOD Matters, US Naval Explosive Ordnance Disposal Facility, Indian Head, Maryland.

DETONATION PROPERTIES OF GUN PROPELLANTS AND SLURRY EXPLOSIVES AND THEIR COMBINATION

H. SCHUBERT
Fraunhofer ICT, Pfinztal, Germany

1. Introduction

The application of demilitarized gun propellants as a component of commercial slurry explosives offer a significant possibility in the demil-business of military energetic materials.

By incineration of gun propellants a large amount of environmental toxic NO_X is produced while by detonation, the produced gases contain only few amount of NO_X. For these purposes aqueous slurry explosives based on ammonium nitrate and additives were filled into propellants containing plastic tubes. After closing the tubes and gelatinizing the material the product is ready for application. The whole process seems to be very simple and economic. Keeping in mind that the environment friendly incineration of propellants will be relative expensive this process may be competitive to the normal explosive production if the demilitarized gun propellants are not charged. Of course the problem of transport licence and permission has to be solved, which is easier in Russia and USA than in Europe. To obtain reproducible detonation-properties of the explosives, the components and their detonation behaviour has to be investigated and experiences should be exchanged.

2. Slurry Explosives

This type of explosives consists of a saturated aqueous solution of ammonium nitrate (65%) and additional amounts of undissolved nitrates in suspension, fuels and gelatinizing agents. Furthermore additives like aluminium powder or microballoons can be used.

The further development is based on a "water-in-oil emulsion" which is formed from a saturated nitrate solution and a mineral oil phase. Microballoons are used to achieve the desired sensitivity.
Slurry Explosives with additional amounts of solid nitrates are nonideal explosives, what means their detonation properties are far below the theoretical values. Depending

on the composition and in consequence on the density, depending on confinement and diameter the detonation rate will be between 3300-4300 m/s. The critical diameter is relatively small and may be decreased by microballoons.

The "Emulsion Explosives" have a higher density and higher performance. Because of the emulsion conditions, the difference between the theoretical and experimental values is smaller (10-15 %). We measure densities between 1,15 -1,35 g/cm^3 and detonation rates between 5000 and 6000 m/s. Critical diameter will be in the range of less than 50 mm and up to 150 mm (without microballoons). A high performance "Emulsion Explosive" with a density of 1,35 g/cm^3 has the following computed values:

Detonation velocity	6613 m/s
CJ pressure	13,95 GPa
CJ temperature	1781 K

3. Solid propellants

There are only few data in literature concerning the detonation behaviour of gun propellants. The main sources are some investigations of deflagration to detonation in powder beds in the frame of low vulnerability and accidents during production or storage of solid propellants. Such accidents were the motivation for investigations later on concerning future safety aspects. Beside other papers especially the Swiss Federal Propellant Plant at Wimmis, TNO, Somchem-Company in South Africa and SNPE have results reported which are useful for our purposes. The information out of literature and my experiences has the following outcome:

The detonation behaviour of solid propellants is influenced by the
- Energy content
- Ambient temperature
- Composition (Module of elasticity, density...)

but also by the

- Shape
- Specific surface
- Porosity and
- Surface treatment
- Perforation of the propellant.

Under distinct conditions also the diameter of the propellant charge, its loading density and the confinement influence the detonation behaviour.

Without confinement the critical diameter of a non-porous extruded doublebase grain is between less than 2 mm to ca. 25 mm depending on the heat of explosion and other influences.

Extensive investigations were done about the deflagration to detonation behaviour of gun propellants by thermal ignition in cylindrical steel cases.

The table 1 shows some results of different gunpowders concerning the critical bulk height by thermal ignition in a confinement. (Wimmis)
Solid propellants - depending on energy content - have theoretical detonation rates between 6500 and
7800 m/s as shown in Table 2 in opposition to a powder bed with smaller rates.
Table 3 shows the max. detonation rates of some gun propellants. The solid propellants in a confinement were ignited by thermal means.

4. Slurry Explosives with gun propellants

Odintsov and Pepekin published in PEP 1996 a paper about "Comparative brisant characteristics of some classes of industrial emulsion explosives". They used in their investigations also gunpowder with the following result:
The emulsion explosive with a density of 1,35 g/cm3 consists of:

Ammoniumnitrate	77 %
Water	16 %
Hydrocarbon	6 %
Oleic acid	1 %

The detonation velocity was 6613 m/s, the CJ pressure 13,95 GPa and the CJ Temperature 1781 K.
By adding gun propellants they found 7432 m/s, 21,76 GPa and 3297 K. All these Batas are computer calculated, therefore the experimental Batas must be much lower.
The type and percentage of gun propellants in the emulsion explosive are not published.

5. Discussion

The aim for the development of gun propellant containing slurry explosive is to improve the reliability and reproducibility of its blasting properties.

To improve the initiation properties of the charges the influence of "Hot Spots" should be investigated. The following questions must be answered:

- Does the perforation of the grain produces "Hot Spots" or are the holes filled up with slurry material during the process
- To which extents wetting of the propellant grains are observed.
- Is it worth to use additional aluminium powder or microballoons to improve the initiation properties of the propellant containing slurries?

- Make it sense to use ammonium nitrate emulsions for propellant containing explosives.

REPRODUCTIVE TECHNOLOGIES OF GUNPOWDER PRODUCTION

F.F.RARAN
Gazizov, GosNIICHP, Kazan, Russia

Associated to the changes in geo-political situation and their influence on the Russian military doctrine a significant part of the Russian army ammunition became to be redundant. In the State reserve was collected a large amount of ammunition, gunpowder and explosives, whose necessity was not more affirmed. Physical conditions of this ammunition evoked the feeling of danger owing to the fact that the guaranty time of its storing was expired. Therefore, actual necessity of solving the utilisation of getting old ammunition, gunpowder and explosives appeared.

At the same time, the process of utilisation is frequently considered as a temporary campaign. Utilisation programmes are connected with political campaigns and are used to solve the ecological and economical problems of armament not by the best way. As a special example, the basic course of nitrocellulose powders utilisation – the preparation of industrial explosives – is unsatisfactory as from the economical point of view, as from the standpoint of technical level of manufactured explosives. However, it is feasible to accent that the operational possibilities of opening this production and the scale of readjusting are unrivalled.

Today, before the entry to the XXI century, it seems to be feasible to pay attention to the process of utilisation as to an inevitable part of the life cycle of the ammunition, destined for assurance of multiple material recycling in military industry sphere. For assurance of such recycling it is necessary to lay basic requirements of reproducibility into the perspective aspects of planning. Here the reproducibility means the possibility of repetitional overwork of moral and physic turning out materials and products to a modern production on contemporary level of technology, design and formulas.

Probably in the nearest future the reproducibility will be the basic criterion of ergonomic efficiency of technologies, underlining not only the production processes without waste, but also the production life cycle without waste generation.

Let us evaluate the problems of reproducibility of Russian technologies. In the present situation in Russia, problems of transformation of gunpowder making to reproductive technologies is complicated by the fact, that the resources of physically getting old gunpowder, storing of which is not more sensible or forms a potential danger, are higher than the raw material needs of the gunpowder industry. Unhappily, there are some formulas, being not prospective according to criteria of reproducibility. In these consequences, new technologies should comply with the requirement of doubled use, i.e. to assure the dividing of gunpowder to separate components and their redoing to requirements of the production for other purposes.

As an example of solving these problems may serve the technology of nitrocellulose regeneration from gunpowder and its readjusting to civilian product based on nitrocellulose or to lacquer basis and lacquer materials on the basic of spherical powders manufacture.

It is clear, that the application of new technologies to transporting-technological conveyers of working production units without sufficient modernisation must not be optimal.

However, too high investment costs are here not warranted, because it is not difficult to manifest that the doubled use of civilian production units is a temporary situation. In consequence with the restriction of gunpowder production, with the sale of nitrocellulose mixtures and civilian powders for export the excess of getting old gunpowder will decrease and the volume of being utilised production will come to terms with the production volumes of new gunpowder. Readjusting of one ton of getting old powder creates following possibilities:
- production of nitrocellulose mixtures produces new turnover in the height of 30 000 roubles per ton;
- acquirement of gunpowder – 50 – 150 000 roubles per ton;
- economy of 4 m^3 of wood species, predominantly coniferous species or of 1 t of cotton;
- economy of about 0,3 – 0,4 t of mineral acid, about 3 000 kW/h of power consumption, about 20 – 25 kcal of heat energy and 500 – 700 m^3 of water consumption
- adequately is cutting back the amount of sewage waters and harmful gases exhalation.

So that introduction of getting old nitrocellulose powders into the resources recycling:

- increases the ecological safety of regions;
- feeds the gunpowder production with a high quality and homogeneous raw material;
- assures the resources economy and lowers the technological load of the environment in the regions where the gunpowder and nitrocellulose industry is located.

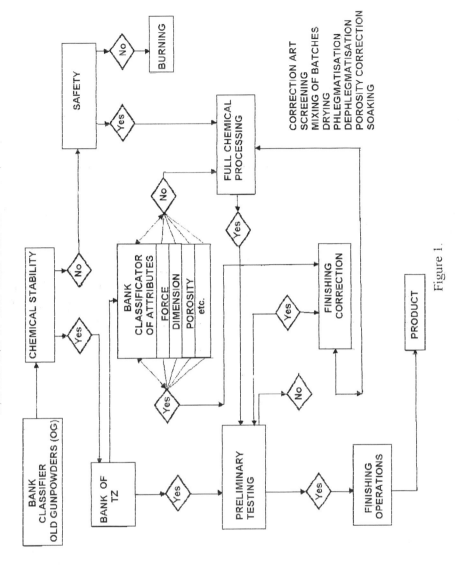

Figure 1.

Let us now treat the technological aspects of gunpowder utilisation in the figures of reproductive technologies of gunpowder production.

In the State Scientific Research Institute of Chemical Products was elaborated an algorithm of gunpowder production on the contemporary level from old gunpowder (see Figure 1). The "inlet" of this algorithm forms a bank – classifier, containing information about the mass, geometrical characteristics, porosity, calorificity and composition of gunpowder elements of being utilised brands and information concerning ballistic characterisation of known military charges, based on these brands.

The "outlet" forms a portfolio of orders for gunpowder of various types, various destiny and for complementary production. Into the basis of the algorithm are included classic technologies of gunpowder making (alcohol–ether extrusive and emulsion technology), adapted for new types of raw materials and new technologies of gunpowder modernisation by partial "insufficient" parameters (shape, porosity, calorificity). These new technologies enable sufficiently reduce the specific costs for the production of new gunpowder types. The modernisation technologies were developed for the needs of civil industry and their realisation needs only limited capital costs.

For technical support of steady market nomenclature of domestic production of gunpowder, both new gunpowder types for domestic firearms and gunpowder types oriented for export purposes, such as cartridge powders for Luger, Remington, Winchester etc., were developed. In the course of development of new generation of gunpowder the criteria of reproducibility of formulas and their ecological safety were satisfied.

The successful activity of the State Scientific Research Institute of Chemical Products in the course of last 10 years in the market of civilian gunpowder's, and a high efficiency of reproductive technologies of gunpowder making grants permission to state the priority of old gunpowder utilisation as a raw material for the production of cellulose nitrates and gunpowder on the concurrent level. The utilization of gunpowder for industrial explosives may be evaluated only as a short-term program for elimination of actual situation. For remake of old gunpowder to industrial explosives may be used predominantly the types of gunpowder, being not appropriate for remake by use of the reproductive technology.

THE USE OF SURPLUS SMOKELESS POWDER PROPELLANTS AS INGREDIENTS IN COMMERCIAL EXPLOSIVE PRODUCTS IN THE UNITED STATES

G. ECK, O. MACHACEK, K. TALLENT
Universal Tech Corporation, Dallas, Texas

1. Introduction

The practice of using surplus propellants as ingredients in commercial explosive products in the United States dates back to the 1950's. They, along with other types of surplus military high explosives, were used as the primary sensitizers in the watergel slurry type blasting agents being developed and sold by the major commercial explosive companies during the late 1950's and early 1960's. Based upon the availability of these surplus smokeless powder propellants, their use by U.S. explosive manufacturers has varied significantly up to the present date. Also, changes in U.S. regulations governing the transportation and storage of commercial explosives has altered the types of explosive products that can utilize propellants as energetic and/or sensitizing ingredients.

In the last six to ten years, a more strict interpretation and enforcement of the US's environmental laws governing the disposition of hazardous waste streams and energetic materials has revitalized the efforts of many U.S. commercial explosive manufacturers to incorporate surplus military explosives and energetic waste streams, generated during the production of military explosives, as ingredients in their commercial explosive products. Because of heavy restrictions being placed upon the often used practices of opening burning and open detonation of these hazardous waste materials, in many instances, the military has been forced to use more costly alternative disposal solutions, such as contained incineration. For these reasons, the U.S. military has encouraged and, oftentimes partially funded the development of technologies based upon the recycling of waste or surplus energetic materials into commercial explosive products. The recycling of such energetic materials as useful ingredients in a commercial explosive does not require the material to be treated as a hazardous waste, according to the U.S. environmental regulations. Furthermore, significant advances in the U.S. military's tactical weapons arsenal has rendered many of its conventional weapon systems obsolete, thereby generating large quantities of similar energetic materials as surplus and in need of disposal. These stockpiles are usually large enough to serve as a reliable and consistent source of raw materials for the development of a commercial explosive product.

2. Background

The subject of this paper, i.e., the use of surplus smokeless powder propellants in U.S. commercial explosive products, dates back to the late 1950's and early 1960's. The main role of these energetic propellants was to provide detonation sensitivity to the product lines of watergel slurry blasting agents that were being developed during this time period. In about 1955, ANFO (acronym for 94 parts ammonium nitrate and 6 parts fuel oil) was developed and immediately began to replace dynamite as the main explosive of choice for most commercial blasting applications. Its ease of manufacture, relatively low cost, handling safety features and good explosive performance properties (energy, VOD etc.) allowed its acceptance in the marketplace to be an overnight success. However, because of ANFO's lack of water resistance and relatively low density, it was not well suited for usage in wet blasting conditions and extremely hard rock applications. Recognizing these shortcomings, the commercial explosive companies of the day developed a number of blasting agent explosives, which came to be known as watergel slurries. These explosive products exhibited the properties of water resistance, densities in the 1.20-1.30 g/cc range and detonation velocities in the 4.5 to 5.5 km/sec range. Although many of these watergel slurry explosive products had energies per unit weight (weight strength), that were significantly lower than that of ANFO, their higher densities allowed them to have higher energies per unit volume (bulk strength) than that of ANFO. A higher bulk strength allowed for more explosive energy to be loaded into a borehole, which produced better fragmentation of the medium being blasted. Furthermore, the watergel slurry's water resistance and density greater than 1.00 g/cc allowed the explosive to be loaded into boreholes containing significant amounts of standing water.

For the most part, the early smokeless powder sensitized watergel slurries were relatively insensitive blasting agents, which were not readily detonated by means of a simple commercial blasting cap of detonating cord. They required a high detonation pressure booster, usually containing a molecular explosive such as cast pentolite or Composition B, in an appreciable mass for reliable detonation. The propellant based blasting agents could be used in either bulk form or as a packaged explosive. The types of explosive products were usually produced in a fixed plant, using suitable mixing and pumping equipment for handling a slurry with a fluid rheology similar to that of oatmeal or fresh concrete. The bulk slurries were pumped directly into the tank of a pump truck, which was then driven to the mine of quarry, where the explosive was pumped through a length of loading hose into the already drilled boreholes, and later detonated. As a packaged explosive, the slurry was pumped into plastic cartridges of varying diameters and weights, as determined by the requirements of its intended field application. The loaded cartridges were either placed into boxes or loaded directly onto a truck for transport to the end user.

From the beginning, explosive engineers clearly understood the advantages of using surplus smokeless powder propellants as inexpensive sensitizers and energetic ingredients in their watergel slurry blasting agents. When subjected to the high pressures generated in the detonation front or primary reaction zone, the burning rate of the propellant increased to the point that it ceased to burn as a propellant, and began to detonate as an explosive. Furthermore, a typical watergel slurry explosive contained water to provide a flowable medium for holding all the explosive's ingredients together in intimate contact allowing them to readily react in the detonation reaction. This aqueous system readily allowed for the use of such water soluble inorganic oxidizer salts as ammonium nitrate, sodium nitrate, ammonium perchlorate and sodium perchlorate, just to mention a few. The entire aqueous system was held together in a fluid viscous gel, created by the addition of a suitable hydrophilic colloid, such as starch or guar gum. The resultant gel rheology could be made fluid enough to be readily flowable and pumpable, while possessing enough viscosity to adequately suspend any undissolved oxidizer salts, insoluble powdered organic or metallic fuels, and the smokeless powder particles. Since the propellant was not soluble in the slurry's aqueous system, it retained its explosive properties. Depending upon the propellant's particle size, it was able to react with the slurry's oxidizer salt to produce increased explosive energy. The presence of water also served to significantly reduce the impact, friction and electrostatic sensitivities of the smokeless powder particles encompassed by the watergel slurry matrix. This was an important feature for achieving the explosive engineer's primary goal of producing a safe explosive product.

3. Explosive Patents

If one examines some of the U.S. patents granted to the commercial explosive companies during the 1960's, there are a number of listings where smokeless powder propellants were used as sensitizing ingredients in commercial explosive formulations. A listing of some of these U.S. patents is given at the end of this text. Although few of these cited early patents actually claim the use of smokeless powder in a commercial explosive formulation, all the patents do describe the use of smokeless powder propellants as a possible sensitizer ingredient in a watergel slurry explosive matrix. For the most part, these slurries contained either single base or double base smokeless powder propellants, although triple base propellants were usually mentioned. Because of the mixing and pumping requirements of the watergel slurry, the small grain 20 mm type smokeless powders were the most popular and easiest to use. However, some companies, Hercules Powder Company for example, installed attrition mills or grinders to downsize the larger propellant grains, that had come out of larger calibre weapons systems, to make them more suitable for use in a watergel slurry explosive product. It was found that smaller propellant grains or particles usually produced a significantly more sensitive blasting agent explosive, with a smaller critical diameter. Therefore, by varying the propellant's type, particle size and use level, an explosive engineer was able to design and produce a particular bulk or packaged blasting agent with the required detonation characteristics (sensitivity, energy, detonation velocity, etc.), suitable for a particular blasting application.

To give an idea of how smokeless powder propellants were used as sensitizing ingredients in the earlier watergel slurry blasting agents, several explosive formulations were taken from the examples listed in some of the cited U.S. patents, identified in the reference section of this paper. These explosive formulations are described in the following sections:

3.1 AMERICAN CYANAMID COMPANY PATENT 3 097 120 (reference listing #1).

This patent gives the following formulation for a smokeless powder sensitized watergel slurry explosive:

TABLE 1.

Ingredients	Weight Percent
Ammonium Nitrate	34.73
Sodium Nitrate	22.58
Smokeless Powder	30.22
Water 12	12.08
Polyacrylamide Thickener	0.36
Aluminium Sulphate	0.03

The smokeless powder used in this slurry was described as being of the 20 mm type, with a formulation of 85% nitrocellulose, 10% dinitrocellulose, and 5% plasticizers and stabilizers. The patent contained detonation test data that clearly showed the slurry explosive's sensitivity to increase as the particle size of the smokeless powder was reduced. The watergel slurry explosive was found to have a critical diameter of 50 mm when the smokeless powder grain size was in the -30/+40 mesh sieve size, and 25 mm when the grain size was 60/+80 mesh.

3.2 IRECO PATENT 3.331,717 (reference listing #12):

This patent describes the use of either single; double or triple base smokeless powders in watergel slurry explosive formulations. The following formulation was presented as an example of a large diameter packaged explosive product:

TABLE 2

Ingredients	Weight Percent
Smokeless Powder	23.9
Ammonium Nitrate	55.8
Water	19.1
Ethylene Glycol	0.8
Guar Gum Thickener	0.4

The smokeless powder used in this slurry was described as having a particle size of 1/32 inch (0.8 mm) by 3/32 inch (2.4 mm). The slurry was found to have a critical diameter of 125 mm at a 1.35 gm/cc density.

3.3 HERCULES POWDER COMPANY PATENT 3,235,423 (reference listing #4):

This patent lists seven different watergel slurry explosive formulations, containing from 25-30% levels of single base smokeless powder as a sensitizer. The smokeless powder grains were described as being either 0.04 inch (1 mm) diameter by 0.1 inch (2.5 mm) long cylindrical rods with a single perforation (1.1 inch single base powder), 0.12 inch (3 mm) diameter by 0.1 inch (2.5 mm) long cylinders with seven perforations (40 mm single base powder), or 0.08 inch (2 mm) diameter by 0.3 inch (7.5 mm) long cylindrical rods with a single perforation (20 mm single base powder). The watergel slurry formulations contained from 15% to 21% water, dual oxidizer salts of ammonium and sodium nitrate, and from 1.5% to 18% ethylene glycol as cold temperature antifreeze. Most of the slurry formulations contained from 14% to 18% granular aluminium as an additional energetic fuel and sensitizer. All the example smokeless powder blasting agent slurry formulations proved to be sensitive enough to detonate in a 75 mm diameter confined charge, with VOD's in the 4.8 to 5.5 km/second ranges.

3.4 HERCULES INCORPORATED PATENT 3,523,048 (reference listing #15):

This patent lists formulations for examples of bulk or pumpable watergel slurry blasting agents, sensitized with smokeless powder. The following slurry formulation was used as an example of suitable bulk watergel slurry explosive:

TABLE 3.

Ingredients	Weight Percent
Water	16.50
Smokeless Powder	32.50
Ammonium Nitrate	38.70
Sodium Nitrate	10.00
Ethylene Glycol	1.50
Guar Gum Thickener	0.70
Fumaric Acid	0.05
Pine Oil	0.05

The smokeless powder used in this bulk watergel slurry formulation was described as having a relatively small particle size, produced by grinding. The resultant watergel slurry had a viscosity and rheology such that it could be readily pumped with a suitable progressive cavity pump.

4. Current Recycling of Surplus Smokeless Powder Propellants in Commercial Explosive Products

As previously mentioned, changes in U.S. environmental regulations, coupled with an increase in the military's munitions demil requirements in the 1990's, have again presented the U.S. commercial explosive companies with a surplus of high explosives and smokeless powder propellants as relatively low cost ingredients for use in their explosive products. Because of changes in the testing requirements for the hazard classification of Class 1 explosive products by the U.S. Department of Transportation for shipping purposes, essentially all of today's smokeless powder sensitized blasting agents are packaged explosive products, and not bulk explosive products.

In order to legalise transport a bulk explosive over the road in the United States, it is necessary for the explosive to be assigned a hazard classification of 1.5. In order to be assigned this hazard classification, the explosive must be subjected to and pass the required test procedures, as described in the United Nations "Recommendations on the Transport of Dangerous Goods Manual of Tests and Criteria". The explosive product must be thermally stable, not impact sensitive, not friction sensitive, and not detonator sensitive. Furthermore, the explosive product must not detonate in an open bonfire test or a deflagration-to-detonation test (DDT). Once the explosive formulation has passed all these required U.N. tests, it can be classified as a Class 1.5 Blasting Agent explosive product, and is approved for transport in an approved package (as described in Chapter 49 of the U.S. Code of Federal Regulations--49CFR). However, if the Class 1.5 Blasting Agent explosive is to be approved for bulk transport in the United States, it must also pass a special semi-confined bonfire test, which was developed by the U.S. Department of Transportation. This U.S. test is referred to as the Large Scale Vented Cookoff Test (or Charlie Schultz test named after the person who developed the test method). The test involves loading 100 pounds (45 kgs.) of the candidate bulk explosive into a 24 inch (610 mm) tall 12 inch (305 mm) diameter schedule 40 steel pipe. The steel pipe container is enclosed on the bottom with a bolted-on 0.5 inch (12.7 mm) thick steel plate. The container's top is enclosed with a welded-on 0.5 inch (12.7 mm) thick steel plate. The steel vessel is vented on top with a 6 inch (150 mm) length of 3 inch (75 mm) diameter schedule 40 steel pipe, welded to the centre of the top steel plate. The loaded steel vessel is placed on top of a three feet tall steel platform, and a suitable wood/fuel oil fire is built under the loaded pipe. The fire must contain enough combustible materials to provide adequate heating for up to 45 minutes. In order to pass this test and be approved for bulk shipment, the explosive cannot detonate during the test. Several explosive companies have made attempts to get smokeless powder sensitized explosive formulations to pass the Large Scale Vented Cookoff Test, without much success. For this reason, most of the surplus smokeless powder propellants are currently being recycled into Class 1.5 packaged explosive products.

5. Current Smokeless Powder Sensitized Blasting Agent Explosive Products

In the current U.S. explosive market, most of the surplus smokeless powder propellants are being recycled into three commercial explosive products. These are all Class 1.5 packaged blasting agents, which are normally marketed in cartridge diameters of three inches and larger. These three commercial explosive products are described in the following sections.

5.1 ORICAI ENERGETIC SOLUTIONS GIANITE™

GIANITE™ can be described as a glass bubble sensitized water-in-oil emulsion packaged explosive, which contains about 35% single base smokeless powder propellant as an energetic ingredient. The emulsion based product usually contains whole propellant grains. These can consist of either cylindrical rods with a single perforation, or larger cylindrical grains with seven perforations (3 to 6 mm diameter X 7 to 13 mm length). The product was developed by ICI Explosives and introduced into the U.S. market in the mid 1990's. The explosive properties and recommended uses for GIANITE™ are detailed in the ICI Explosives published product data sheet attached to the end of this paper. The product is marketed and sold as a packaged booster sensitive explosive, in cartridge diameters of 100 mm and larger.

5.2 DYNO NOBEL DYNOGEL™ HD

DYNOGEL™ HD can be described as a smokeless powder sensitized watergel slurry packaged blasting agent. The actual development work on this explosive product has been described in a paper presented by Dyno Nobel and Olin personnel at the 2nd Global Demilitarization Symposium, held in Arlington, Virginia, in May, 1994. As described in this paper, DYNOGEL™ HD was originally developed to use Olin's "Pit Powder" as a sensitizer in an ammonium nitrate / sodium nitrate based watergel slurry matrix. This "Pit Powder" was described as a water-wet granular double base by-product propellant, resulting as a waste from Olin's ball powder sporting ammunition production. The paper presented the following typical formulation for the "Pit Powder" propellant:

TABLE 4.

Ingredients	Weight Percent
Nitrocellulose	89.7
Nitroglycerine	3.9
Dibuthylphtalate	3.6
Ethyl Centralite	0.2
DPA and derivatives	1.4
Dinitrotoluene	0.5
Ash	0.6
Sand	0.1

The paper described DYNOGEL™ HD as containing about 35% of this "Pit Powder" propellant as its primary sensitizer and fuel. The watergel explosive was described as having a 2.5 inch (64 mm) critical diameter, a minimum booster of 4.5 grams, a minimum temperature of -20°C, a density of 1.4 g/cc, a detonation velocity of 5.3 km/sec and a critical pressure of 160 psig. (1100 kPa). Underwater energy test data showed the DYNOGEL™ HD to generate a Shock Energy of 371 cal/gm and a Bubble Energy of 428 cal/gm, for a Total Energy of 799 cal/gm (as compared to a Shock Energy of 332 cal/gm, a Bubble Energy of 548 cal/gm and a Total Energy of 880 cal/gm for standard ANFO in the same test). On a theoretical basis, DYNOGEL™ HD was described as having a Relative Weight Strength of 0.91 (ANFO = 1) and a Relative Bulk Strength of 1.41 (ANFO = 1).

The paper also contained a schematic of a production flow diagram for the DYNOGEL™ HD explosive product. The watergel slurry was mixed in a ribbon blender, pumped into a holding bin to await packaging, and pumped into 75 mm diameter and larger flexible polywoven cartridges or shotbags.

Since its original development, Dyno Nobel has expanded the types of smokeless powder propellants that are used to produce DYNOGEL™ HD. For the most part, they are currently using whole grain surplus and demifled single and double base propellants. These can vary from ball powders to one or seven perforation propellant grains. Attached to the end of this text is a copy of Dyno Nobel's published Technical Information sheet for the DYNOGEL™ HD product. As can be seen, Dyno Nobel markets this booster sensitive blasting agent in cartridge diameters of three inches and larger. The explosive is promoted as a high density, high velocity, high shock energy, and „deadpress" resistant blasting agent, suitable for use in extreme blasting conditions.

6. Slurry Explosives Corporation Slurran 430

Slurran 430 can be described as a smokeless powder sensitized watergel slurry packaged blasting agent. The Slurran 430 product differs from the previously mentioned packaged propellant explosives, in that it contains a significantly higher percentage of smokeless powder propellant. The Slurran 430 product contains about 60% whole grain triple base smokeless powder propellant. The triple base propellant grains normally have a cylindrical or hexagon shape, and usually contain either 7 or 19 perforations. The propellant grains can vary in diameter from 7 mm to 12 mm and in length from 12 mm to 26 mm.

The Slurran 430 product was developed for Slurry Explosives Corporation (SEC) by Universal Tech Corporation (UTeC) personnel. The actual development of this product has been described in a paper presented by UTeC personnel at the 6th Global Demilitarization Symposium, held in Coeur d' Alene, Idaho, in May of 1998. As described in this paper, it was found that relatively large grain smokeless powder propellants could be made to readily detonate in their original grain configurations, when surrounded by a high density liquid medium. During the development work, it was determined that the higher the density of this liquid medium, the easier it was to make the propellant detonate and the smaller the resultant explosive's critical diameter. The following table lists some of the detonation data generated with 8 mm diameter X 16 mm long single base (M6 formulation) and triple base (M30 formulation) 7 perforation grains:

TABLE 5.

Propellant/Medium	Unconf. Charge Dia.	Detonation Results
M6 / air	150 mm	Failed to Detonate
M6 / water	150 mm	Failed to Detonate
M6 /gelled salt soln.	150 mm	VOD = 6.2 km/sec
M6 /gelled salt soln.	100 mm	VOD = 6.0 km/sec
M6 /gelled salt soln.	75 mm	VOD = 5.8 km/sec
M6 /gelled salt soln.	50 mm	VOD = 5.5 km/sec
M6 /gelled salt soln.	38 mm	Failed to Detonate
Propellant/Medium	**Unconf. Charge Dia.**	**Detonation Results**
M30 / air	150 mm	Failed to detonate
M30 / water	150 mm	VOD = 6.6 km/sec
M30 / water	100 mm	VOD = 6.5 km/sec
M30 / water	75 mm	VOD = 6.4 km/sec
M30 / water	50 mm	VOD = 5.9 km/sec
Propellant/Medium	**Unconf. Charge Dia.**	**Detonation Results**
M30 / water	38 mm	Failed to Detonate
M30 /gelled salt soln.	150 mm	VOD = 7.0 km/sec
M30 /gelled salt soln.	100 mm	VOD = 6.9 km/sec
M30 /gelled salt soln.	75 mm	VOD = 6.8 km/sec
M30 /gelled salt soln	50 mm	VOD = 6.8km/sec
M30 /gelled salt soln.	38 mm	VOD = 6.5 km/sec

Note: The gelled salt solution consisted of an aqueous solution of ammonium nitrate, sodium nitrate and calcium nitrate, gelled with guar gum, which had a density of 1.5 g/cc. All charges were primed with a 0.45 kg cast pentolite booster and detonated at 20°C.

As can be seen from the above test data, both types of smokeless powder propellant produced a fairly sensitive explosive product, with a detonation velocity in the 5.8 to 6.8 km/sec range, in a 75 mm diameter unconfined charge, when surrounded with the high density aqueous salt solution. In addition to the Slurran 430 product, SEC has started production of a Slurran 406 product, which is made with single base whole grain smokeless powder propellant.

The explosive energies of the Slurran 406 and Slurran~430 products were measured using the underwater energy test. These energy data are expressed in the following table, on both a unit weight basis and a unit volume basis. For comparative purposes, ANFO was included as a standard explosive in these tests.

TABLE 6.

Explosive	Density	Shock	Bubble	Total
ANFO	0.90 m/cc	360 cal/m	520 cal/m	880 cal/m
ANFO	0.90 m/cc	324 cal/cc	468 cal/cc	792 cal/cc
Slurran 406	1.45 m/cc	454 cal/m	436 cal/m	890 cal/m
Slurran 406	1.45 m/cc	658 cal/cc	632 cal/cc	1290 cal/cc
Slurran 430	1.50 m/cc	482 cal/m	409 cal/m	891 cal/m
Slurran 430	1.50 gm/cc	732 cal/cc	614 cal/cc	1346 cal/cc

As would be expected, the triple base propellant slurry produced a higher Shock Energy component, while the single base propellant produced a higher Bubble Energy component. On a unit weight basis, both propellant slurries produced a significantly higher Shock Energy component than ANFO, while producing a lower Bubble Energy component. Overall, on a unit weight basis, both propellant slurries produced about the sameTotal Energy as that of ANFO. However, with the higher densities of the propellant slurries, their bulk energies were significantly higher than those of ANFO.

Because of the high percentage of smokeless powder propellant in the Slurran 406/430 products, they cannot be produced using the traditional mixing and pumping equipment for a typical watergel slurry explosive. As described in the UTeC paper, the production process for the Slurran 406/430 products is quite simple. It involves first filling a polywoven shotbag with the required volume or weight of propellant grains. Then, a mixture of a delayed gelling solution and the oxidizer salt solution is poured over the column of propellant until all the grains are submerged. Then the cartridge is clipped shut and loaded into a box or directly onto a truck for transport. The gelling agents eventually thicken the product's liquid phase to the extent that it resembles gelatine. Because of the high percentage of propellant grains in the final explosive product, the cartridges are rigid and do not slump well in the borehole. Also, because of the relative insensitivity of the propellant sensitized watergel slurry explosive, the product requires good loading practices, such as intimate bag-to-bag coupling and primer-to-bag coupling in the borehole, as well as the use of a high detonation pressure cast booster, to achieve consistent performance.

Attached to this report is a copy of SEC's technical data sheet, which describes the use advantages and properties of the Slurran 430 product. As can be seen, SEC markets the Class 1.5 blasting agent in cartridge diameters of 75 mm and larger. Because the Slurran 430 product contains smokeless powder as its only sensitizer, it is promoted as being a "deadpress" resistant explosive.

7. Conclusions

The use of conventional ammunition propellants as ingredients in commercial explosives has in the past, and still continues today; to offer a viable method for the recycling of demilitarized smokeless powder propellants. This technique not only allows for the beneficial use of the surplus energetic materials as an effective ingredient in commercial explosive products, but it also solves many environmental and economic problems associated with their potential destruction as a hazardous waste.

REFERENCES

General References:

1. Bonner, C.D., Cranney, D.H., Funk, A.G., Drummond, J.A., (1994) Commercial Blasting Agents Using Surplus Propellants, *Proceedings 2nd Global Demilitarization Symposium, Meeting #472, pp. 265-280, May 1994.*

2. Eck, G.R., Tallent, K.D. (1998), Common Test Methods Used To Measure Properties Of Commercial Explosive", *Proceedings of B.A.I. Eighth High Tech Seminar, pp 435-459. July 1998.*

3. Machacek, O., Eck, G.R., Tallent, K.D., (1998), Development of New High Energy Blasting Products Using Demilitarized & Excess Propellant Grains from Conventional Ammunition Rounds, *Proceedings of 6th Global Demilitarization Symposium & Exhibition, May 1998.*

4. *United Nations, Recommendations on the Transport of Dangerous Goods Manual of Tests and Criteria--Second Revised Edition, 1995.*

5. *ICI Explosives Technical Data Sheet No. 218 (2M/8-96) for GIANITE^{TM}.*

6. Dyno Nobel, Inc., *Technical Information Sheet (P-13-4-13-94) for DYNOGEL^{TM} HD.*

7. Slurry Explosives Corp., *Technical Information Sheet for Slurran 430.*

United States Patents:

1. Hoffman, J.A., Bowkley, H.L., American Cyanamid Company, (1963,) *No. 3,097,120*, July 9, Gelled Ammonium Nitrate Explosive Containing Polyacrylamide And An Inorganic Crosslinking Agent

2. Bowkley, H.L., Merryweather, J.P., American Cyanamid Company, (1963), *No. 3,097,121*, July 9, Powdered Ammonium Nitrate Explosive Containing Polyacrylamide And An Inorganic Crosslinking Agent.

3. Ursenbach, W.O., Udy, L.L., IRECO, Inc., (1993) ,*No. 3,113,059*, Dec. 3, Inhibited Aluminium-Water Composition and Method.

4. Fergusion, J.D., Hercules Powder Company, (1966). *No. 3,235,423*, Feb. 15, Stabilized Aqueous Slurry Blasting Agent And Process.

5. Ferguson, J.D., Hopler, R.B., Hercules Incorporated, (1966), *No. 3,288,658*, Nov. 29, Aerated Explosive Compositions.

6. Swisstack, P.L., Hercules Incorporated, (1966), *No. 3,288,611*, Nov. 29, Aerated Aqueous Explosive Composition with Surfactant.

7. Lyerly, W.M., E.I. duPont de Nemours and Company, (1997), *No. 3,355,336*, Nov. 28, , Thickened Water-Bearing Inorganic Oxidizer Salt Explosive Containing Crosslinked Galactomannan And Polyacrylamide.

8. Chrisp, J.D., E.I. duPont de Nemours and Company, (1967), *No. 3,297,502*, January 10, Explosive Composition Containing Coated Metallic Fuel.

9. Cook M.A., Pack, D.H., Gardner, J.N., IRECO, Inc., No. 3,331,717, Inorganic Oxidizer Blasting Slurry Containing Smokeless Powder and Aluminium.

10. Cook, M.A., Clay, R.B., (1968) *No. 3,371,606*, March 5, , Explosive Booster For Relatively Insensitive Explosives.

11. Cook, M.A., IRECO, Inc., (1968) *No. 3,379,587*, April 23, Inorganic Oxidizer Salt Blasting Slurry Composition Containing Formamide.

12. Cook, M.A., IRECO, Inc., (1968) *No. 3,382,117*, May 7, Thickened Aqueous Explosive Composition Containing Entrapped Gas.

13. Albert, A.A., Hercules Incorporated, (1968) *No. 3,390,031*, June 25, Gelled Aqueous Slurry Explosive Composition Containing An Inorganic Nitrate.

14. Jessop, H.A., Udy, L.L., IRECO Inc. (1969) *No. 3,485,686*, Dec.23, Aqueous Explosive Slurry Containing Oxidizer-Reducer Crosslinking Agent.

15. Hopler, R.B., Hercules Incorporated, (1970) *No. 3,523,048*, August 4, , Bulk Delivery of Crosslinkable Aqueous Slurry Explosive With Crosslinking Agent In A Separate Feed.

16. Machacek, O., Eck, G.R., Gilion, J.B., (1997) *No. 5,608,184*, March 4, Alternative Use of Military Propellants as Novel Blasting Agents.

BURNING AND DETONATION OF WATER-IMPREGNATED COMPOUNDS CONTAINING SOLID AND LIQUID PROPELLANTS

B.N. KONDRIKOV, V.E. ANNIKOV AND V.YU. EGORSHEV
Mendeleev University of Chemical Technology, Moscow, Russia

Introduction

Utilization of propellants and explosives extracted froth ammunition for rock blasting and some other purposes is a new kind of chemical technology originated after the cold war was ended, eventually in the end of the 80ies or even beginning of the 90ies. Correspondingly the new scientific description of the new field of industry is needed. In this work we are going to present some physical and chemical data concerning mechanisms of burning and detonation of the water-impregnated systems. Owing to the great diversity of the systems, specific requirements of the industry to these compounds and in many cases of a big difference between these new and traditional explosive materials, investigation of the mechanisms represents specific direction of combustion science. Eventually, this paper comprises a brief review of the investigations carried out in Mendeleev University of Chemical Technology, Moscow, during the last two decades and also, during two or three last years, in Politecnico di Milano. They are described partially in the works [1-16].

Ten types of the systems are considered in [1-16].

1. Water-impregnated explosive compounds (WEC) on the base of ammonium and sodium nitrate containing powdered aluminium, in some of the compounds up to 90%, for electric-hydrodynamic and other installations.
2. Aquanals sensitized with pigment grade aluminium powder and non-explosive organic fuel, say carbamide.
3. Aquanals containing an active organic fuel, methylamine nitrate in addition to aluminium.
4. Waterproof (water-resistant) compounds of the water-in-oil type containing, as a combustible, a mixture of petroleum products (industrial oil, heavy oil or crude oil) with an emulsifier providing the combustible distribution in the format of a thin film on the surface of the finest oxidizer suspension droplets.
5. WEC containing RDX and HMX;

6. WEC on the base of single and double base propellants
7. Compositions on the base of inorganic salts of perchloric acid
8. Systems on the base of organic salts of perchloric acid.
9. WEC containing HAN and

The last kind of WEC have no industrial application so far but due to the ingredients of liquid (HAN) propellant are/were produced industrially and corresponding technologies of manufacturing of these compounds still exist, the practical ground for their scientific study is quite evident. One may suggest also some novel areas of commercial application of these substances, say in form of the sort of gelled propellants because they possess of ability to steady state combustion with well-defined and regulated speed [11,13,15]. Thus combustion and detonation behaviour of the substances would be of interest not only immediately but also potentially.

Experimental

In order to prevent separation of the compounds in the storage process, the salt-water solution was gelatinized with polyacrylamide (1.0-2.5%. pulverised granulated product). The crosslinking of polyacrylamide in the gelled solution was performed with potassium dichromate in the presence of sodium thiosulfate. Additives insoluble in water were introduced in the matrix before cross-linking. Water-in-oil compositions were prepared by means of a gradual addition of the concentrated AN/SN water solution having a temperature of approximately 90 °C to the mixture of the fuels, i.e. crude oil, or industrial oil, and emulsifier, at a very intensive stirring. The substance of a consistence of a dense cream (at viscosity of about several hundreds Poises) was produced. Burning experiments were performed in a constant pressure bomb (up to about 40 MPa) with a registration of the process on the quickly moving film. 20-μm-thick tungsten-rhenium (80/20 and 95/5) micro-thermocouples were used to measure temperature profiles. Copper-constantan thermocouples with the bead size of 50 μm, coated with a thin layer of glass, were employed to measure the combustion temperature of HAN solution. The liquids to be tested were poured into 7-8 mm, sometimes 12 mm, id. polycarbonate or quartz tubes. All the detonation experiments glass tubes of different diameter or steel tubes of 36/10 mm diameter and 200 mm long were used.

Results

BURNING.

WEC of the types 1-5, 9 and 10 were investigated. Some of the data are presented in Table 1 and Figure 1 to 3 (At the presentation, essentially bigger number of Tables and Figures will be given).

The main results of the investigation are as follows.

1. The PDL, (Pressure Deflagration Limit) and BRL. (Burning Rate Limit at PDI) are very high numbers at burning of almost all of the WEC systems (see Table 1). One may conclude correspondingly that WEC at a moderate pressure is characterized by the lack of stability.

2. At a high pressure (30 MPa) the burning stability WEC is greater. The burning rate, at least for the gelled substances burning at the expense of gas phase reactions is close to the values for usual well-known nitrocompounds [17] (Fig.2). In the WEC on the base of pigment grade aluminium powder (Compounds # 3 to 6 in Table 1)

TABLE 1 Compositions and some characteristics of water impregnated compounds investigated

No	Components content, %				T_6, K	ΔP, MPa	v	PDL MPa	R_b, mm/s at a pressure (MPa)		
	H_2O	AN/SN	PG/CP	Fuel					30	10	PDL
1	15,0	33,0/12,0	0/40	-	3970	1-36	0,8	1,1	6	2,4	0,5
2	15,0	33,0/12,0	4,0/36,0	-	3970	1-36	0,5	0,6	6	3,4	0,7
3	20,0	45,0/15,0	10,0/0	10,0 U	2253	20-36	1,2	20	13	-	6
4	18,0	40,0/13,0	20,0/0	9,0 U	2943	6-18 18-36	1,5 0,0	6,0	31	13,6	6
5	15,0	35,0/12,0	30,0/0	8,0 U	3323	2-9 9-36	1,5 0,3	2,0	33	23,7	2
6	10,0	44,0/17,0	4,0/5,0	20,0 M	2489	2-36	0,8	2,0	17	7,1	2
7	11,0	48,0/18,0	-	23,0 M	1579	6-36	1,2	4,0	13	3,5	0,6
8	20,0	28,0/9,0	-	43,0 M	1488	5-16	1,5	4,0	-	3,5	0,8
9	10,0	34,0/10,0	5,0/0	41,0 R	2844	0,4-36	1,0	0,4	26	9,1	0,5
10	15,0	60,0/18,0	-	7,0 C	1793			>36	-	-	-
11	20,0	38,0/0	-	42,0 M	1472	1,8	1,8	20	12	-	6
12	22,0	49,0/17,0	-	11,0 U	1040	-	-	36	-	-	5
13	19,0	42,5/14,0	15,0/0	9,5 U	2442	1,7 0,7	1,7 0,7	14	21	-	8
14	19,0	42,5/14,0	15,0/0	9,5 U	2442	1,6	1,6	12	19	-	5
15	19,0	42,5/14,0	15,0/0	9,5 U	2442	1,6 0,7	1,6 0,7	13	16	-	6
16	14,5	61,0/14,5	-	10,0 C	1523	0,8	0,8	16	5	-	3
17	14,2	59,6/14,2	-	12,0 C	1377	1,0	1,0	14	5	-	2
18	13,7	57,6/13,7	-	15,0 C	1273	0,9	0,9	21	5	-	3
19	13,2	55,6/13,2	-	18,0 C	1227	1,1	1,1	30	3	-	3
20	12,2	61,1/17,7	-	9,0 I	1742	1,4	1,4	26	4	-	4
21	14,2	43,4/10,4	-	12,0 C/ 20,0 M	1291	2,2	2,2	11	14	-	1,5

Notes: The pressures are given in MPa, U is carbamide (Urea), M is Methylamine nitrate, R is RDX; I is Industrial oil (the percentage includes an emulsifier), C is Crude oil, PG is Pigment Grade aluminium powder, CP is a Crude Powder. The compositions 14 and 15 are almost identical to composition 13, with the exceptions that 14 does not contain $Na_2S_2O_3$ and 15 does not contain $K_2Cr_2O_7$. DP is interval of pressures studied

The interesting peculiarity of burning can be noted: The flame temperature is determined with participation of aluminium in the burning reactions, whereas rate of the leading reaction at this temperature is connected with interaction of combustibles and oxidizers in flame.

3. At the moderate pressure and for the substances characterized by the strong contribution of reactions in condensed phase (WEC #7, 8, 11, 16-19 in Table 1), the burning rates are much higher than those estimated on the base of the experiments leading to the straight line in Figure 2. Calculation of the kinetic constants of the leading reactions in condensed phase can be implemented in suggestion that surface temperature is determined by the boiling point of AN/water solution. Calculation thus performed gives activation energy of the reaction in AN/MAN mixture 28.5 kcal/mol and of the reaction of aluminium oxidation 17 kcal/mol. Both are coincide with the numbers derived from the usual kinetic experiments. It would be relevant to compare the last value with the figure obtained in [13] for powdered Magnesium oxidation at burning in a mixture with AN and potassium dichromate. It is only 6 kcal/mol; quite naturally it is much less than the activation energy of oxidation of aluminium is.

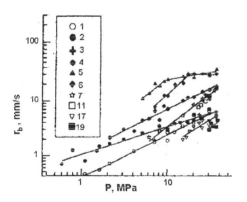

Fig. 1.. Burning rate vs pressure of some of the water-impregnated compositions listed in Table I.

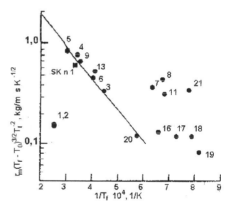

Fig. 2. Dependence of the reduced burning rate of water-impregnated materials at p = 30 MPa vs calculated inverse temperature of flame. Straight line corresponds to the dependence for nitrocompounds [17].

(Numbers near points or curves in both Figures correspond to the data in Table I)

4. Transition from moderate to high pressure is followed sometimes by the mesa u(p) curve formation (see, curves for the substances 4 and 5 in Table 1 and Fig. 1). The surface temperature increases slightly in this area of pressures, reaction in the condensed phase is stabilized or goes on to the diffusion regime. Meanwhile the gas - phase reactions are not fast enough to be responsible for this value of burning rate. Correspondingly, the burning rate augmentation is stopped till the gas phase reactions become fast enough to produce the further enhancement of the burning rate with pressure.

5. The leading role of the convective heat exchange is manifested at burning of HAN containing WEC. Influence of the exponent v in Vielle burning law is demonstrated. At $v < 0.5$ the convective burning rate versus pressure becomes anomalous dependence, the rate diminishes at pressure rise. It elucidates the fact of burning rate vs pressure decrease in the interval of pressures 8 to 30 MPa. Moreover, at 20-23 MPa the Landau limiting burning rate responsible for transition from conductive to convective burning regime becomes less than the real conductive burning rate. Correspondingly the sharp decrease of the rats in this pressure interval can be also explained.

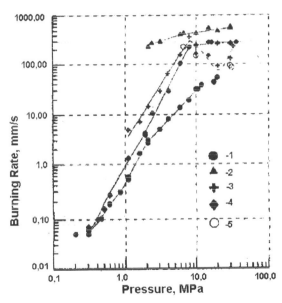

Fig.3. Effect of pressure on the burning rate of crystalline HAN (1), liquid propellant 57.5% HAN/37.5%, EAN/5% H_2O (2), gelled and cross-linked 64% HAN solution in water (3), gelled liquid propellant 57.5% HAN/37,5% EAN/5% HBO (4), and 64% HAN solution in water (5).

6. Burning of solid crystalline HAN was first studied (Figure 3). The consequence of chemical reactions to explain the main peculiarities of HAN burning was proposed. The kinetic constants of the overall reactions are $E = 26.5$ kcal/mol, $\log_{10} Z\ (s^{-1}) = 14.2$.

DETONATION.

Experimental results on detonation of WEC of the types of 1 to 8 are described earlier [1-9, 12, 14,16]. The new experimental results with WEC of the types 9 and 10 are given in Table 2.

The main conclusions concerning mechanism of detonation are as follows.

I .All of the WEC with the exception of the compounds of

the types 5 and 6, even in the heavy walled steel tubes, in the absence of the air bubbles preliminary introduced into the mass of explosive detonated in the low velocity detonation regime (LVD). The limiting explosion heat corresponding to the border of detonation failure/LVD is 2.5⁄2.7 MJ/kg. In paper tubes the failure diameter is noticeable greater, 6/7 mm. The failure diameter of HAN/NMA/water solutions is still more great, 14.8/18.7 mm. It relates to the boundary between extinction of detonation and LVD. It should be stressed again that HVD could not be observed in these systems.

TABLE 2. Detonation of HAN, MAN, AN water solutions in the steel tubes

Content, %				ρ	Qv	Result	D,
HAN	MAN	AN	Water	G/ml	MJ/kg		Km/s
87,1	0	0	12,9	1,51	0,87	Detonation	
81,3	0	0	18,7	1,49	0,66	Extinguish	
48,5	44,4	0	7,1	1,42	4,14	Detonation	
48,5	44,3	0	7,2	1,42	4,13	Detonation	
48,5	44,3	0	7,2	1,42	4,13	Detonation	2,55
40,6	37,1	0	22,3	1,34	3,09	Detonation	
37,9	34,7	0	27,4	1,29	2,74	Extinguish	
18,7	39,8	38,7	2,8	1,44	4,08	Detonation	2,50
23,5	48,5	24,5	3,5	1,43	4,11	Detonation	2,50
32,8	42,0	20,3	4,9	1,43	4,11	Detonation	
32,8	42,0	20,3	4,9	1,43	4,11	Detonation	2,34
23,9	38,0	27,6	10,5	1,40	2,71	Detonation	
23,1	36,8	26,6	13,5	1,35	3,45	Detonation	2,43
28,1	35,9	17,4	18,6	1,37	3,18	Detonation	
21,7	34,6	25,1	18,6	1,36	3,11	Detonation	
20,3	32,3	23,4	24,0	1,33	2,75	Detonation	
19,4	31,4	22,4	26,8	1,27	2,59	Extinguish	
18,9	30,2	21,9	29,0	1,30	2,42	Extinguish	
18,5	29,5	21,3	30,6	1,31	2,31	Extinguish	

2. WEC containing the big concentration of RDX, HMX or single Vase propellant are able to detonate in the HVD regime in a broad region of high explosives or propellant content. The limiting concentration of thin dispersed (4μm) RDX in concentrated solution of AN in water is as small as 15/20% in the steel tubes. It is 25/30% at the same sort of matrix in the glass tubes of 18-19 mm id., and 30/35% in the tubes of 14-15 mm id. The limiting concentration of HMX (250 μm) in the matrices with no active fuel is 40/45 % and in the matrix containing MAN is 35/40%.

3. Essentially different result was observed in experiments with WEC containing single base propellant for a gun of a small calibre (seven channels 3 mm diameter grain, 4 mm long). In the oxidizing matrix without combustible ingredients (with the exception of very small content of polyacrylamide, which is not considered a fuel in the common sense) it detonates poorly at a concentration of the propellant only 57%, (three detonations and two cases of extinction of detonation in five experiments). At the same

time in the matrix which does contain fuel, both MAN and Carbamide, it detonates even at 35% of the propellant (exactly the same result, 3 detonation and 2 extinction cases).
4. WEC containing dispersed double base propellant (0,4 mm-mean particle size) did not produce true HVD in any matrix at any concentration. Even at 50 and 60 % of the propellant in the mixture. Detonation, in the bubble sensitized LVD regime though, was observed in the matrix on the base of MAN at a concentration of the propellant of 20% (diameter of the tube was 18 mm, detonation velocity 2.4 km/s) al density of the mass only 0.74 of TMD. One may note however that this matrix may detonate at this density with no propellant addition.

Acknowledgement
Authors wish to thank Russian Foundation for Basic Research for the partial financial help of the investigations (Grants 32164 and 32167)

References

1. Annikov, V.E,. Borzykh, M.N., Kondrikov, B.N., (1978) Detonation of water-gel explosives, *Detonation, Akad. Sci. USSR, Chernogolovka*, 19-23.
2. Annikov, V.E,. Borzykh, M.N., Kondrikov, B.N., Korneev, S.A., (1979) Detonation of water solutions of perchloric acid salts, Chemical Physics of Condensed Explosive Systems, *Proceedings of Mendeleev Institute of Chemical Technology, Moscow*, **104**, 95-98.
3. Annikov, V.E., Kondrikov, B.N., Korneev, S.A., (1980) Detonation of water-gel explosive compounds on the base of sodium perchlorate, Problems of Theory of Condensed Explosive Systems, *Proceedings of Mendeleev Institute of Chemical Technology, Moscow*, **112**, 106-112.
4. Annikov, V.E., Kapustin, P.P., Kondrikov, B.N., (1980) Effect of pressure on detonation ability of water-gel explosive containing aluminium., Problems of Theory of Condensed Explosive Systems, *Proceedings of Mendeleev Institute of Chemical Technology, Moscow*, **112**, 101-106.
5. Annikov, V.E., Kondrikov, B.N., Korneev, S.A., Smagin, N.P. (1982) Detonation and combustion of solutions on the base of sodium perchlorate, *Vzryvnoe delo*, No.84/41, *Moscow, Nedra*, 38-42.
6. Annikov, V.E., Kondrikov, B.N., Kapustin, P.P., (1982) Effect of initial temperature on detonation ability of water-gel explosives, *Vzryvnoe delo*, No.84/41, *Moscow, Nedra*, 35-38.

7. Annikov, V.E., Kondrikov, B.N., Korneeva, N.N., Puzyrev, S.N., (1983) Detonation mechanism of gas-filled water-gels, *Comb., Expl. & Shock Waves*, **19**, No.4, 139-143.
8. Andriyanova, E.E., Annikov, V.E., Kondrikov, B.N., Korneeva, N.N. (1985) The structure of detonation wave of aerated watergel mixtures, *3-th All-Union Seminar on Detonation, Tallinn*, , 20-21.
9. Kondrikov, B.N., Tarutin, V.P., Annikov, V.E. (1988) Kinetic peculiarities of detonation of water solutions of perchlorates, *Comb., Expl. & Shock Waves*, **24**, No 3, 75-80.
10. Kondrikov, B.N., Andriyanova, E.E., Annikov, V.E. (1988) Recondensation of viscous foamed liquids, *Proceedings of Akad. Sci. USSR (Doklady)*, **298**, No 6, 1433-1436.
11. Egorshev, V.Yu., Kondrikov, B.N., Yakovleva, O.I. (1991) Combustion of water-impregnated explosive compounds, *Comb., Expl. & Shock Waves*, **27**, No.5, 56-64.
12. Alymova, Ya.V., Annikov, V.E., Kondrikov, B.N. (1994) Detonation of the emulsion explosive, *Proc. of 20th Int. Pyrotechnics Seminar*, Colorado Springs, Colorado, USA, 11-23.
13. Kondrikov, B.N., Annikov, V.E., Egorshev, V.Yu., DeLuca, L, Bronzi, C., (1999) Combustion of ammonium nitrate - based compositions, *Journal of Propulsion and Power*, in press.
14. Annikov, .E.V. Kondrikov, .B.N., Kazakov, A.T., (1998) Research and development of the new explosive sources of seismic waves for geophysical investigations, *11th Detonation Symposium, Snowmass, Colorado*, Paper Summaries, 370372. Should be published in the Symposium Proceedings.
15. Kondrikov, B.N., Annikov, V.E., Egorshev, V.Yu., DeLuca, L, (2000) Burning of hydroxylammonium nitrate, *Comb., Expl. & Shock Waves*, 2000, **36**, in press.
16. Kondrikov, B.N., Annikov, V.E. (1997) Detonation of water-impregnated compositions containing propellants and high explosives, *NATO Workshop, Socorro, NM*, April 1997.
17. Kondrikov, B.N., Raykova, V.M., Samsonov, B.S. (1973) Kinetics of the combustion of nitrocompounds at high pressure, *Comb., Expl. & Shock Waves*, **9**, No. 1, 84-90.

CHARACTERIZATION OF INTERMOLECULAR EXPLOSIVES

R. M. DOHERTY
Naval Surface Warfare Center, Indian Head, MD,
C.O. LEIBER
*Wehrwissenschaftliches Institut für Werk-, Explosiv-, and Betriebsstoffe
Swisttal, Federal Republic of Germany*

Abstract: The use of recovered rocket and gun propellants in commercial slurry explosives will require some understanding of the characteristics of intermolecular explosives, since the fuel-oxidizer mixtures that are common in propellants will, when made detonable, constitute explosives of this type. The slow energy release and strong dependence of initiability on properties of the mixture such as porosity and intimacy of mixture of the components must be taken into account when the sensitivity and performance of these materials is assessed.

1. Introduction

The possibility of converting gun and rocket propellants that have been removed from munitions into a useful source of energy for other purposes is very appealing. It not only offers a means of avoiding the generation of air pollutants that arise from open burn or open detonation, but also allows the recovered energetic materials to be put to some productive use as an energy source. Among the options for reuse of demilitarized propellants in this way are use of the energetic materials mixed with other fuels in boilers, and the utilization of the recovered propellants in commercial explosives. The latter is the topic of this workshop.

The nature of the propellants that are likely to be recovered from rockets is very similar to the class of explosives that may be designated intermolecular explosives. For the purposes of this paper, intermolecular explosives (IMEs) will be defined as those explosives that derive a large fraction of their total energy from reactions between components of an explosive mixture, as opposed to coming from the decomposition of a molecular explosive such as RDX, HMX, or TNT. IMEs are well known and very widely used in the field of commercial explosives, but the ageing characteristics of many IMEs make them impractical for use in military applications.

IMEs are intrinsically heterogeneous. Since, by definition, much of the energy comes from reactions between ingredients in the explosive composition, there must be more than one component. Typically the main ingredients may be classified as fuels and oxidizers. One of the simplest IMEs, in terms of composition, is ammonium nitrate-fuel oil (ANFO). In this case the ammonium nitrate is the oxidizer and the fuel oil is the fuel. Other examples of IMEs are shown in Table 1[1].

TABLE 1. Compositions of Some Intermolecular Explosives

Name	Fuel	Oxidizer
Tovex A-4	TNT; Al	AN; SN
EAK	EDDN	AN; KN
Amonit Skalny I H	Wood meal; petroleum tar	AN
SE (slurry explosive)	Al	SP
No.5 Kuro Carlit	Ferro-silicon, sawdust	AP

AN = ammonium nitrate; SN = sodium nitrate;
KN potassium nitrate; EDDN = ethylenediamine dinitrate;
SP = sodium perchlorate; AP = ammonium perchlorate

The most common types of oxidizers used in commercial explosives are nitrate salts. Ammonium nitrate in particular is inexpensive and readily available. The nitrate salts also have an environmental advantage over perchlorate salts (ammonium perchlorate (AP) is commonly used in propellants) because they do not generate HCl in the products of combustion or detonation. Other components can also act as the oxidizer, provided they carry sufficient oxygen relative to the amount of fuel they bear. One example of this is nitroglycerin, which has a positive oxygen balance.

The fuels that can be used in IMEs are to be found in two main classes: carbonaceous materials and metals. Carbonaceous fuel that have been used in commercial explosives run the gamut from materials that are themselves energetic, such as TNT and nitrocellulose, to biomass materials such as wood meal, peat, flour and plant fibres. Lubricating oil and mineral oil have also been used as fuels. Polymeric materials, such as the polybutadienes used in rocket propellants, are also good fuels. The primary metal fuel that has been used is aluminium. It has a high heat of oxidation, is readily available in a variety of forms (flake, and particles of sizes from a few microns to hundreds of microns), and is relatively inexpensive.

In addition to these main components of commercial IMEs, other ingredients may be added, depending on the use to which the material will be put. Explosives used in coal mines are formulated to minimize the chance of accidental dust or methane explosions.

Thus they omen contain salts such as sodium, potassium, or ammonium chloride, to reduce the detonation temperature and suppress explosions involving air and dust or methane. The general insensitivity of many commercial explosives in the absence of porosity (see below) prompts the addition of materials that can act as sensitizers. Flake aluminium or explosive components such as TNT or nitroglycerine can be added to increase the sensitivity of the explosive, as well as to increase the total energy of the composition. Glass microballoons are also an effective means of sensitizing IMEs. Further ingredients may be added to stabilize the composition chemically (compounds to neutral acid or absorb impurities), to stabilize the composition mechanically (thickening agents), or to increase the intimacy of mixture of the fuel and oxidizer (water in emulsion explosives, surfactants or emulsifiers).

2. Chemical reactions in IMES

The nature of the chemical reactions in IMEs determines the behaviour of these materials. Rather than being driven by a monomolecular decomposition or the reaction of like molecules separated only by the spacing within a crystal lattice or a liquid, IMEs by definition depend upon two different species, the fuel and the oxidizer, coming together in order to react and thus liberate energy. The cases in which the best intimacy of mixture of the fuel and oxidizer is achieved are the ones that most nearly approximate the efficiency of energy delivery that is associated with monomolecular explosives.

The case of eutectic salts based on ethylenediamine dinitrate (EDDN) and ammonium nitrate (AN) is illustrative. Hershkowitz and Akst [2] have reported that the method of preparation of a mixture of these two components influences the performance. Although EDDN is an explosive with energy similar to TNT, its performance can be improved when an oxidizer such as is AN available to oxidize the extra carbon in EDDN. The extent of the performance improvement depends on the method of mixing the two components. The more intimately the components were mixed, the better was the performance. By bringing the oxidizer and fuel into closer contact, the time required for reaction can be reduced.

3. Price's group 1 and group 2 explosives

Price [3] distinguished two groups of high explosives based on the shape of the dependence of their detonation velocities on porosity. The main division was between pure explosives and mixtures of explosives with a nonexplosive fuel. The trend for most high explosives, designated Group 1 explosive, is for the detonation velocity to increase linearly with increasing percentage of theoretical maximum density (TMD), and for the lines describing such this relationship for cylindrical charges of differing diameters to converge to a common point as he density approaches TMD as is shown in Figure 1. This type of behaviour is exhibited by most pure high explosives, though there are some exceptions, such as nitroguanidine.

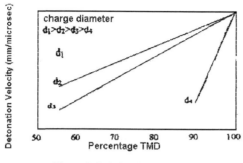

Figure 1. Relationship of Detonation Velocity to % TMD

In contrast to the Group 1 behaviour, explosives that are dominated by fuel-oxidizer for Group 1 Explosives reactions, i.e., IMEs, show a decreasing detonation velocity as %TMD increases. At relatively low %TMD, the same trend of increasing detonation velocity with increasing %TMD is observed, but as the porosity is removed for the charge, the exothermic reactions that support the propagation of the detonation wave are quenched, and the detonation velocity slows down. Eventually the explosive fails to propagate a steady detonation. This type of behaviour is shown by AP as shown in Figure 2.

The behaviour of AP also illustrates another characteristic of Group 2 explosives: the change in detonability limit with porosity. As the TMD increases (porosity decreases), the detonation velocity first increases, then drops off; at some point the charge fails. As the charge diameter increases, the %TMD at which a detonation can be propagated in the charge also increases, and the performance, as measured by the detonation velocity, approaches the maximum attainable velocity, as calculated by an equilibrium code such as TIGER or CHEETAH.

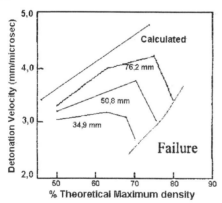

Figure 2. Dependence of AP Detonation Velocity on %TMD (Ref. 2)

The Group 2-type dependence of failure diameter on density has been shown by Lee et al. [4] for a water-in-oil emulsion explosive comprising AN/water/oil/emulsifier in a ratio of 77/16/6/1. They observed an increase in the failure diameter as a function of density. Although the explosive could be detonated in a diameter of less than 25 mm at a density of 1.2 g/cm [3] (89% TMD), the failure diameter increased to over 225 mm as the density approached TMD. These charges were very lightly confined, in PVC pipe, but it has been suggested that even this confinement may have influenced the failure diameter.

The behaviour observed for the Group 2 explosives is a result of the time required for the relevant chemical reactions to take place, and the effect of temperature on the reaction rates. In Group 1 explosives, the main reactions take place under conditions of high temperature and pressure, and the bulk of the reaction occurs near the detonation front. In Group 2 explosives the temperatures and pressures at the detonation front are generally lower than in Group 1 explosives, and surface reactions are prominent, even in conditions of steady state detonation. The pure explosives that exhibit Group 2 behaviour tend to have low detonation temperatures. It is easier to see the importance of the surface reactions in IMEs, in which reactions between the components can occur only where they are mixed.

4. Initiation of IMEs

The discussion above is centred on the propagation of detonation in IMEs, but the initiability of this type of material is extricable tied to the same properties that confer Group 2 behaviour on IMEs. The question of initiability is very important, not only because it affects choices that are made regarding the reliable intentional initiation of IMEs and their total energy output, but also because it influences decisions made regarding the safety of IMEs. The classification of explosives as Extremely Insensitive Detonating Substances (EIDS) is based upon the response of an explosive in a test that is very similar to the Expanded Large Scale Gap Test (ELSGT) [5]. In this test, as in all common gap tests, the size of the explosive is fined, and the response of the explosive to shocks of various levels is monitored. The intensity of the shock introduced is diminished from the shock of the donor material (typically Composition B, pentolite or something similar) by a variable length gap of cellulose acetate or poly(methyl methacrylate), PMMA. In the ELSGT, the diameter of the charge is 73.2 mm, and it is encased in a steel tube of 11 mm wall thickness. The length to diameter ratio of the charge is 3.8 (27.9 cm long). In order for an explosive to be classified as an EIDS, which affects how it may be transported and stored, there must be no detonation (as judged from a steel witness plate at the end of the charge) with a gap of 70 mm. This corresponds to an input pressure of about 35 kbars. Even this may not be sufficiently large to give an accurate assessment of the hazards associated with large quantities of IMEs.

A smaller gap test than the ELSGT is commonly used to distinguish between Hazard Class/Division 1.1 (mass detonating) and 1.3 (mass deflagrating) materials. The NOL Large Scale Gap Test (LSGT) [6] employs an explosive sample that is 36.5 mm in diameter, confined by a steel sleeve with 5.5 mm wall thickness. The length of the sample is only 140 mm, for a length to diameter ratio of 3.8, as in the ELSGT.

A problem arises from the fact that the response of an explosive in a gap test depends on the scale of the test relative to the failure diameter of the charge, as well as the distance required for a low level reaction to run up to a detonation. Even for some compositions that show Group 1 behaviour, the level of shock required to initiate a detonation has been shown to depend on the size of the charge at sizes that are many multiples of the failure diameter. The Super Gap Test [7] was devised to assess the hazards associated with the storage of large quantities of explosives. The results of this test for showed that, in the diameter of this test (8"), the shock required to detonate Composition B was 12 kbar, compared to 18 to 20 kbar in the NOL LSGT. Tritonal, an explosive with lower oxygen balance and a non-explosive fuel, also shows a significant difference between its sensitivity in the LSGT (25 kbar) and its sensitivity in the Super Gap Test (15 kbar).

IMEs are known to show behaviour in both performance and sensitivity tests that changes with charge diameter [2,4,8]. Since many of the common tests used to characterize explosives were originally designed for compositions with generally smaller failure diameters and shorter reaction zone lengths, it is worth remembering that "Experiments designed for ideal explosives are not always appropriate for nonideal explosives." [9]. The problem of the applicability of small scale tests to IMEs involves not only the scale of the experiment but the lack of the definitive transitions that are ordinarily observed with Group 1 explosives. This is reflected in the relatively flat slope of the plot of charge initiation pressure vs. run distance to detonation for two eutectic salt mixtures, EAK 46 (46/46/6) and EAK SO (50/42/8) [8]. Akst noted that it was not possible to fit a true Pop-plot to the data, owing to the lack of clear-cut transitions. However, since a detonation could be initiated in a sufficiently long run distance by input pressures comparable to those required for Composition B, Akst stated that "...these materials might be said to exhibit low explosiveness - likelihood of escalation to very violent reactions - but not low sensitiveness." Similar behaviour was observed by Dick [10] in his study of the initiability of a rocket propellant.

5. Modelling of the behaviour of IMEs

The modelling of IMEs has lagged behind that of more ideal explosives due to the difficulty of treating late reactions that take place behind the sonic plane. Though the energy released in these can still contribute to mechanical work, it is often difficult to calculate *a priori* what the performance of these materials will be. Recent developments in the incorporation of a kinetic model into a performance prediction code has made it possible to improve the predictions, but the practice is to characterize the output of IMEs by test methods that take into account energy release of a long time. Such techniques include the measurement of underwater bubble energy and lead block expansion (Trauzl test) [1].

6. Reuse of reclaimed propellants

TABLE 2. Compositions of some rocket and gun propellants [1]

Propellant	Fuel	Oxidizer	Other
Space Shuttle Booster	PBAN binder, Al	AP	curing agent, catalyst
JPL X-500	PPG ginder	AP	crosslinker, catalyst
Single/double base propellants	Nitroglycerin, Nitrocellulose		burn rate modifiers, stabilizers

The utility of reclaimed rocket and gun propellants in commercial explosives will depend on many factors, only some of that are related to the energetic properties of the materials. The economics of developing a method of using a product of which there is only a limited supply works against this. The effort to be expended in characterizing an intrinsically heterogeneous material manufactured using recovered material of uncertain history will be very large. In determining the sensitivity and the performance of commercial explosives made with recovered propellants, the usual precautions must be taken to ensure that factors such as the uniformity of distribution of components that are intended to react with one another, uniformity of sensitizers such as microballoons, and suitability of the test conditions are taken into account.

The compositions of some representative rocket and gun propellants are given in Table 2. As can be seen, one of the common ingredients in rocket propellants is AP. This may cause some environmental problems if it is incorporated into commercial explosives due to the HCl that is one of the products of AP decomposition. Propellants typically also contain burn rate modifiers, which may be lead-based or contain other heavy metals. Although the concentrations of such components are small, consideration should be given to their potential environmental impact.

7. Conclusion

Since slurry explosives belong the class of explosives known as Intermolecular Explosives, they may be expected to exhibit the characteristics of this class of materials, including intrinsic heterogeneity. The assessment of the utility of recovered rocket or gun propellants in slurry explosives, or other types of IMEs, must recognize the importance of the characteristics of IMEs to the measurement of sensitivity, safety, and performance.

REFERENCES

1. Urbanski, T. (1984) *Chemistry and Technology of Explosives, Volume* 4, Pergamon Press, Oxford.
2. Hershkowitz, J, Akst, I. (1976) Improvement of Performance of Composite Explosives Containing Ammonium Nitrate by Physical Synthesis, *Sixth Symposium (international) on Detonation, White Oak, MD*, ACR-221, p. 439.
3. Price, D. Contrasting Patterns in the Behaviour of High Explosives, *Eleventh Symposium (International) on Combustion, Berkeley, CA*, p. 693, 1966.
4. Lee, J et al.,(1989) *Ninth Symposium (International) on Detonation, Portland, OR*,
5. Liddiard, T.P, Price, D. (1987) *The Expanded Large Scale Gap Test*, NSWC TR 86-32, March 1987.
6. Price, D., Clairmont, A.R., Jr. and Erkman, J.O. (1974) The NOL Large Scale Gap Test, III, Compilation of Unclassified Data and Supplementary Information for Interpretation of Results, *NOLTR* 74-40, March 1974.
7. Foster, J.C., Forties, K.R., Gunger, M.E. and Craig, B. G. (1985) An Eight-Inch Diameter, Heavily Confined Card Gap Test, *Eighth Symposium (International) on Detonation, Albuquerque, NM*, p. 228.
8. Akst, I.B. Intermolecular Explosives, (1985) *Eighth Symposium (International) on Detonation, Albuquerque, NM*, p. 1001.
9. Kennedy D. Comment in Reference 4.
10. Dick, J.J. (1981) Nonideal Detonation and Initiation Behaviour of a Composite Solid Rocket Propellant, *Seventh Symposium (International) on Detonation, Annapolis, MD, NSWC MP 82234*, p. 620.

SAFETY ASPECTS OF SLURRY EXPLOSIVES

NICO H.A. van HAM
TNO Prins Maurits Laboratory
Rijswijk, Netherlands

Summary

The Royal Dutch Navy in co-operation with TNO has developed a method to remove the explosives from munitions by means of high-pressure water jets. After separation of the explosives from the metal parts, munitions components can be reused. The TNT filling can be used as a component for industrial explosives, and the metal shells can be reused as practise rounds.

Explosives that result from the dismantling of munitions cannot be reused in all cases, and must be disposed consequently. A possible way of disposing explosive waste is by incinerating of the explosives. During the incinerating of explosives there always will be the risk of an explosion. Due to the characteristics of the explosive, the physical and chemical reactions that might occur, a (slow) burning reaction might turn into a violent incineration or even a detonation. This will lead to loss of life and property.

The most common way to suppress accidental reactions in explosives is to moisten the explosives with water. A mixture of at least 30 % water by weight is no longer classified as explosive. In order to prevent the settlement of the explosive in the water, additives are used resulting in stable slurry that can be stored for at least several months.

To evaluate the possible risks during the handling and incineration of slurry, safety test were performed on slurry consisting of 50 % TNT and 50 % water. These tests included the sensitiveness towards impact, friction, electrostatic discharge and heat. The same tests were repeated with the dried slurry to investigate the possible risk of slurry spilling and subsequent drying of the slurry.

Next step is to study the reaction of the slurry under confinement. This confinement may occur in the process equipment. Due to the confinement explosives usually show very high reaction rates and pressures. The slurry in the process equipment can be ignited by external heat sources, such as fire and short-circuiting. Tests include the Koenen test and the Deflagration to Detonation Transition (DDT) test.

Most severe test is the detonation test. In this test the ability of the slurry to show a detonating reaction can be detected.

The tests that were performed clearly showed that it was very difficult to ignite the slurry, once ignited no propagating reaction could be detected. Even a very strong shock wave from an adjacent detonating explosive dissipated in the slurry resulted in the fading out of a propagating shock wave.

It was concluded that TNT slurry is safe to handle, safe to transport and safe to store.

1. Introduction

The main characteristic of explosives is the ability to generate enormous amounts of energy in a time frame of microseconds. This power can be used for military and civil applications, that cannot be realised by other means. The drawback of these uncanny forces is the risk for accidental reactions during production, handling, transportation, storage, use and finally disposal. Accidental reactions may lead to loss of life and properties.

To suppress the start of unwanted reactions, a common procedure in the explosives industry is to moisten the explosives with water, or otherwise dilute the explosives with inertisating agents, to desensitise the explosive. This is frequently practised during production and transportation of explosives. This safety feature can also be applied in the disassembly of munitions

Together with the Netherlands Royal Navy, TNO has developed a water washout procedure, to remove the explosives from munitions that are no longer needed. This is a safe method to separate the explosives from the metal parts of the munitions. The explosives can subsequently be reused e.g. as industrial explosives. If the reuse is not possible, the disposal of the explosive should be realised as soon as possible to eliminate the risk of accidental reactions. The most attractive way is by incinerating the mixture of the explosive and the water. In order to prevent the settling of the explosive in the water, stabilisers are added to create stable slurry.

Beside the approved safety of the whole process, much attention was paid to the labour hygienic aspects and the environmental aspects. There is no TNT dust. There is very little TNT vapour, the part of the building where the TNT is removed from the shells is strictly separated from the rest of the building. All the water necessary in the water wash out process is in a closed system, which is in use for almost 10 years, so there is no wastewater problem. These items were already discussed during the Advanced Workshop in Moscow in May 1994.

In this study the sensitivity of slurry consisting of TNT from munitions and water is investigated. The different test methods simulate the different risks for accidental reactions in the industrial environment. The whole process of the water washout, the slurry preparation, the slurry feed and incineration of slurry in a Fluidised Bed Oven (FBO) was patented by TNO [1].

2. Definition of test methods.

Given the nature of the slurry, test methods should be aimed at establishing the thermal and explosive properties of the mixture. Also the sensitivity to mechanical action and electrostatic discharge is of importance to evaluate the risk.

2.1. DIFFERENTIAL SCANNING CALORIMETER (DSC), THERMO GRAVIMETER (TG) AND DIFFERENTIAL THERMAL ANALYSIS (DTA)

These tests give an indication of the behaviour of the material during heating; e.g. decomposition energy and loss of mass / material in a certain time.

2.2. VACUUM STABILITY TEST (VST)

In this test, the thermal stability of the slurry mixture is determined. This is of importance for the long time storage of the slurry.

2.3. SENSITIVENESS TO IMPACT AND FRICTION

These test methods determines the sensitiveness of the mixture to friction and impact, and thus the probability and risk for the initiation of undesired reactions that might occur during handling and transport of the slurry.

2.4. DETONATION TEST

This test determines whether the slurry is capable of propagating a detonation.

2.5. DEFLAGRATION TO DETONATION TRANSITION (DDT)

This test determines whether a deflagration (explosive burning) of the slurry can result in a detonation.

2.6. HEATING UNDER CONFINEMENT (KOENEN TEST)

In this test, the effect of heating the slurry under confinement is determined

2.7. TIME PRESSURE TEST

In this test, the effect of ignition of the slurry under confinement is determined.

2.8. CAP SENSITIVITY TEST

In this test the sensitiveness of the slurry to intense mechanical stimuli is determined.

2.9. SENSITIVITY TO ELECTRO STATIC DISCHARGE (ESD)

In this test, the reaction of the (dried) slurry to an electrostatic discharge is determined.

The results of the above mentioned tests can be used to assess the hazards and risks of the disposal process of the slurry. The sensitivity to impact, friction and other mechanical stimuli gives insight in the likelihood of an undesired event. With the result of the detonation test, Koenen test, time pressure test and DDT test the maximum effect in case of an incident can be predicted. The data obtained from DSC, TG, DTA and VST can be used to better understand the reactions occurring. The result of the ESD test can be used to assess the effect of sparks on the substance. Although the water containing slurry will be insensitive to sparks, there will be a (remote) possibility that the slurry might dry during an incident; in this case the results allows to predict the effect.

3. Test results

3.1 THERMAL TESTS

3.1.1 *DSC*

With the Differential Scanning Calorimeter it can be easily determined which endothermic and exothermic effects occur within a certain temperature limit. The results can also be used to decide whether a more sensitive measuring technique should be used to determine the thermal stability in relation to production, storage and transport. The DSC apparatus used (SCIKO DSC-220-C) is capable of investigating a temperature range of 130 to 1 000 K. The temperature of the sample under test is compared with a reference substance. In this way, the net thermal effect of the sample is measured. Since only a small sample is tested, the inhomogeneity of the sample may lead to relatively large variations.

The result of the DSC test is shown in Figure 1. This result is comparable to the result of pure TNT indicating that the water and the stabiliser added to improve the slurry's properties do not influence the decomposition behaviour of the TNT.

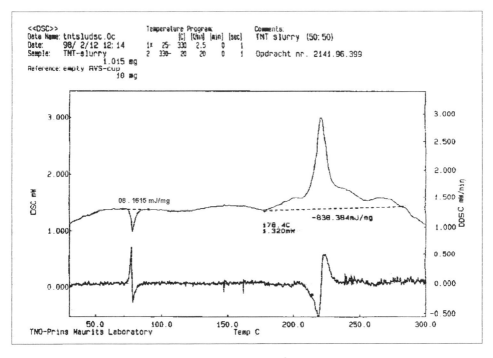

Figure 1. DSC plot of TNT slurry

3.1.2. *TG/DTA*

In an unstable substance, or an unstable mixture of substances, several chemical and physical phenomena may take place, such as: reduction - oxidation reactions (redox reactions), decomposition reactions, evaporation and water separation. Those reactions are temperature dependent and are characterised by changes in mass and energy. One of the most logic combinations to study these reactions is with thermogravimetry and differential thermal analysis.

Thermogravimetry is the determination of the mass change of a sample as a function of temperature (scanning) or time (isothermal). Continuous mass registration is obtained with a thermo-balance. The sample is placed in a horizontal oven, which can be kept at a constant temperature or can be heated with different rates between 0.01 and 100°C/min in the range between ambient temperature and 1500°C.

An example of a combined thermogram is given in Figure 2.

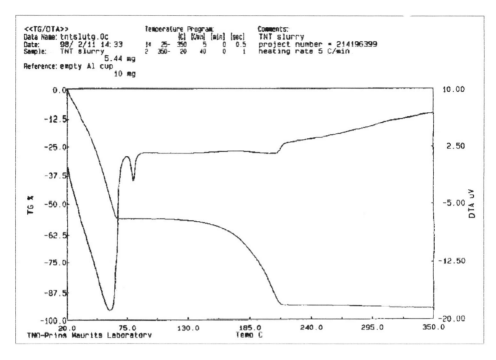

Figure 2: Thermogram with TG and ETA signal.

The thermo-balance (Figure 3) consists of a DTA cell, which is placed on the arm of the balance. The measuring cell is confined in an oven made of aluminiumoxide. The oven is protected against external temperature influences by a metal reflector.

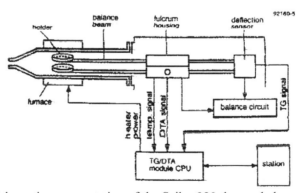

Figure 3: Schematic representation of the Seiko-320 thermo balance. (92160-5)

Besides the possibility of measuring in a certain atmosphere (air, nitrogen, oxygen, helium) there is also the possibility to record a thermogram in vacuum.

The experiments are carried out in open vessels, in the temperature range of 25 to 375° C with a heating rate of 5°C/minute. There is an endothermic peak at 75° C; this is the melting of the TNT. A mass decrease of about 50% was measured at 70° C, resulting from the evaporation of water. In the temperature range between 125 and 230° C, there was a further decrease of 44 mass-%. These results confirm the DSC conclusion (3.1.1) that there is no other decomposition reactions in the slurry mixture than those characteristics of TNT.

3.2. VACUUM STABILITY TEST

With the vacuum stability test the amount of gas evolved in a certain time at a certain temperature in vacuum is measured. The amount of gas is a measure for the decomposition rate of a substance. This test is standardised between NATO countries and is described in [3]. The test is usually applied to propellants and secondary explosives, mostly at temperatures of 120 °C, 100 °C and 90 °C, depending on the substance. Normally, samples of 5 grams of material are used at a test time of 40 hours. Other test conditions may be employed as well.

In this case, a temperature of 100 °C was selected. During a 40-hour period 2.92 ml and 2.78 ml of gas per 2.5 g of test substance was measured. Considering the amount of water in the slurry the result is satisfactory.

3.3. SENSITIVENESS TO IMPACT AND FRICTION

3.3.1. *Sensitiveness to impact*

TNO uses the BAM fallhammer. This is the international standard apparatus to test the impact sensitiveness developed by the Bundes Anstalt fur Materialprüfung (BAM). The essential parts of the fallhammer (see Fig.4) used to assess the sensitiveness are a cast steel block with base, an anvil, a column, guides, drop weights with release device and an impact device. A steel anvil is screwed onto the steel block and cast base. The support, into which is fixed the column (made from a seamless drawn steel tube), is bolted to the back of the steel block. The dimensions of the anvil, the steel block, the base and the column are given in Figure 4. The two guides which are fixed to the column by means of three cross-pieces are fitted with a toothed rack to limit the rebound of the drop weight and a movable graduated scale for adjusting the height of the fall. The drop weight release mechanism is adjustable between the guides and is clamped to them by the operation of a lever nut on two jaws. The apparatus is fixed onto a concrete block (600 x 600 mm) by means of four anchoring screws secured in the concrete, so that the base is in contact with the concrete over its whole area and the guides are exactly vertical. A wooden protective box with inner protective lining and which can be opened easily, surrounds the apparatus up to the level of the bottom crossbar. An extraction system allows removal of any explosion gases or dust from the box.

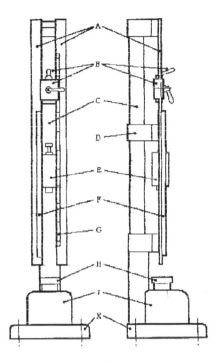

Figure 4. BAM Fallhammer: General view

A	Two guides	F	Toothed rack
B	Holding and releasing device	G	Graduated scale
C	Column	H	Anvil 100 mm diameter x 70 mm
D	Middle cross-piece	J	Steel block 230 x 250 x 200 mm
E	Drop-weight	K	Base 450 x 450 x 60 mm

The drop weight is shown in Figure 5. Each drop weight is provided with two locating grooves holding it between the guides as it drops, a suspension spigot, a removable cylindrical striking head and a rebound catch which are screwed on to the drop weight. The striking head is of hardened steel (HRC hardness of 60 to 63); its minimum diameter is 25 mm; it has a shoulder piece preventing it from being forced into the drop weight by the impact. Three drop weights are available with the following masses, 1.00 kg, 5.00 kg and 10.00 kg. The 1 kg-drop weight has a heavy steel centre fitted with the striking head. The 5 kg and 10 kg drop weights are of massive and compact steel, e.g. material specification at least St 37-1 in accordance with DIN 1700.

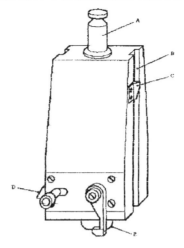

Figure 5. Drop weight

A	Suspension spigot
B	Positioning groove
C	Height marker
D	Rebound catch
E	Cylindrical striking head

A sample of the substance under test is enclosed in an impact device consisting of two co-axial steel cylinders, one above the other in a hollow cylindrical steel guide ring. The cylinders are steel rollers from roller bearings with polished surfaces and rounded edges and HRC hardness between 58 and 65. The impact device is placed on an intermediate anvil and centred by a locating ring with a ring of vent-holes to permit the escape of gases.

The cylinders and the guide ring were degreased with acetone before use. The cylinders and guide ring are only be used once.

For substances in powdered form, a sample is taken with a cylindrical measure of 40 mm^3 capacity (3.7 mm diameter x 3.7 mm). For paste-like or gel-type substances, a cylindrical tube of the same capacity is inserted into the substance and after levelling off the surplus, the sample is removed from the tube by means of a wooden rod. For liquid substances, a fine-drawn pipette of 40 mm^3 capacity is used. The substance is placed in the open impact device, which is already in the locating ring on the intermediate anvil, and for powders or paste-like or gel-type substances, the upper steel cylinder is gently pressed until it touches the sample without flattening it. Liquid samples are placed in the open impact device in such a way that it fills the groove between the lower steel cylinder and the guide ring. The upper steel cylinder is lowered, with the aid of the depth gauge, until it is 2 mm from the lower cylinder and held in place by a rubber "O" ring. In some cases, capillary action causes the sample to exude from around the top of the sleeve. In these cases, the assembly should be cleaned and the sample re-applied. The filled impact device is placed centrally on the main anvil, the protective wooden box is closed and the appropriate drop weight, suspended at the required height, is released. In the interpretation of the results of the trial, distinction is made between "no reaction", "decomposition" (without flame or explosion) recognisable by change of colour or odour and "explosion" (with weak to strong report or inflammation). In some cases it is advisable to perform trials with appropriate inert reference substances to allow a better judgement of whether or not an audible report has occurred.

The limiting impact energy, characterising the impact sensitiveness of a substance, is defined as the lowest impact energy at which the result "explosion" is obtained from at least one out of at least six trials. The impact energy used is calculated from the mass of the drop weight and the fall height (e.g. 1 kg x 0.5 m = 5 J). The 1 kg drop weight is used at fall heights of 10 cm, 20 cm, 30 cm, 40 cm and 50 cm (impact energy 1 to 5 J). The 5 kg drop weight is used for fall heights of 15 cm, 20 cm, 30 cm, 40 cm, 50 cm and 60 cm (impact energy 7.5 to 30 J) and the 10 kg drop weight for fall heights of 35 cm, 40 cm and 50 cm (impact energy 35 to 50 J). The series of trials is started with a single trial at 10 J. If at this trial the result "explosion" is observed, the series is continued with trials at stepwise lower impact energies until the result "decomposition" or "no reaction" is observed. At this impact energy level, the trial is repeated up to the total number of six if no "explosion" occurs; otherwise the impact energy is reduced in steps until the limiting impact energy is determined. If at the impact energy level of 10 J the result "decomposition" or "no reaction" (i.e. no explosion) was observed, the test series is continued by trials at stepwise increased impact energies until for the first time the result "explosion" is obtained. Now the impact energy is lowered again until the limiting impact energy is determined.

Tests were performed with the slurry, dried slurry, dried 'emulsifier' and dried TNT. In the first three cases an impact sensitivity of greater than 50 Nm was found. In the latter case a sensitivity of 4 Nm was measured. For comparison: in the United Nations Manual on transport regulations [2] a substance is too sensitive to be transported if the impact sensitivity is equal to or smaller than 2 Nm. From these results it can be concluded that the slurry is not sensitive to impact. Even in the case the slurry might dry, the dried slurry is far less sensitive compared to pure TNT.

3.3.2. Sensitiveness to friction
The friction apparatus (see Figure 6) consists of a cast steel base, on which the friction device is mounted properly. This comprises a fixed porcelain peg and a moving porcelain plate (see next paragraph). The porcelain plate is held in a carriage, which

runs in two guides. The carriage is connected to an electric motor via a connecting rod, an eccentric cam and suitable gearing such that the porcelain plate is moved, once only, backwards and forwards beneath the porcelain peg a distance of 10 mm. The loading device pivots on an axis so that the porcelain peg can be changed; it is extended by a loading arm, which is fitted with 6 notches for the attachment of a weight. Zero load is obtained by adjusting a counterweight. When the loading device is lowered onto the porcelain plate, the longitudinal axis of the porcelain peg is perpendicular to the plate. Different weights of masses up to 10 kg are used. The loading arm is fitted with 6 notches of distances of 11 cm, 16 cm, 21 cm, 26 cm, 31 cm and 36 cm from the axis of the porcelain peg. A weight is hung into a notch on the loading arm by means of a ring and hook. The use of different weights in different notches results in loads on the peg of 5 N, 10 N, 20N, 40 N, 60 N, 80 N, 120 N, 160 N, 240 N and 360 N. If necessary, intermediate loads may be used.

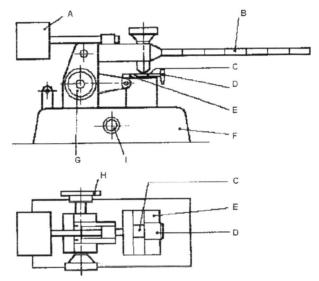

Figure 6. BAM Friction apparatus.

A	Counter weight	F	Steel base
B	Loading arm	G	Handle for setting the carriage in the starting position
C	Porcelain plate held on carriage	H	Points the direction to electric motor drive
D	Adjusting rod	J	Switch to activate motor
E	Movable carriage		

The flat porcelain plates are made from technical white porcelain and, before being fired, their two rubbing surfaces (roughness 9 - 32 microns) are thoroughly roughened by being rubbed with a sponge. The sponge marks are clearly visible. The cylindrical porcelain pegs are also in technical white porcelain and their roughened ends are rounded. The dimensions of plate and peg are given in Figure 7.

Normally the substances are tested as received. Wetted substances should be tested with the minimum content of wetting agent.

Each part of the surface of the plate and peg must only be used once; the two ends of each peg will serve for two trials and the two friction surfaces of a plate will each serve for three trials.

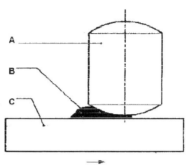

Figure 7. Porcelain plate and peg.

A	Porcelain peg mm diameter x 15mm
B	Sample under test
C	Porcelain plate 25 x 25 x 5 mm

A porcelain plate is fixed on the carriage of the friction apparatus so that the grooves of the sponge marks on it run transversely to the direction of movement. The quantity to be tested, about 10 mm^3 was taken from substances in powdered form by means of a cylindrical measure (2.3 mm diameter x 2.4 mm deep); for paste-like or gel-type substances, a rectangular 0.5 mm thick gauge with a 2 x 10 mm window is used; the window is filled with the substance to be tested on the plate, and the gauge is removed carefully. The firmly clamped porcelain peg is placed onto the sample as in Figure 7, the loading arm is loaded with the required weights and the switch is operated. Care must be taken to ensure that the peg rests on the sample, and that there is enough of the substance to come under the peg when the porcelain plate moves in front of the peg.

The series of trials was started with a single trial at a load of 360 N. The results of each trial are interpreted in terms of "no reaction", "decomposition" (change of colour or odour) or "explosion" (report, crackling, sparking or flame). If in the first trial the result "explosion" is observed, the series is continued with trials at stepwise lower loads until the result "decomposition" or "no reaction" is observed. At this friction load level the trial is repeated up to the total number of six if no "explosion" occurs; otherwise the friction load is reduced in steps until the lowest load is determined at which no "explosion" occurs in six trials. If in the first trial at 360 N the result "decomposition" or "no reaction" occurs, up to five further trials are performed. If in all six trials at the highest load, the result "decomposition" or "no reaction" occurs, the substance is deemed to be insensitive to friction. If an "explosion" is obtained, the load is reduced as above. The limiting load is defined as the lowest load at which the result "explosion" is obtained from at least one out of at least six trials.

The test results are assessed on the basis of:
- whether an "explosion" occurs in any of up to six trials at a particular friction load;
- the lowest friction load at which at least one "explosion" occurs in six trials.

The test result is considered "positive" if the lowest friction load at which one "explosion" occurs in six trials is less than 80 N and the substance is considered too dangerous for transport in the form in which it was tested. Otherwise, the test result is considered "negative".

Tests were performed with the slurry, the dried slurry, the dried 'emulsifier' and dried TNT. In all cases a friction sensitivity of greater than 360 N is found. For comparison: in the transport regulations [2] a substance is too sensitive to be transported if the friction sensitivity is equal to or smaller than 80 N. From these results it can be concluded that the slurry or even the dried slurry is not sensitive to friction.

3.4. DETONATION TEST

This test is used to measure the ability of a substance to propagate a detonation by subjecting it to a detonating booster charge under confinement in a steel tube.

The apparatus is shown in Figure 8 and is identical for solids and liquids. The test sample is contained in cold-drawn, seamless, carbon steel tube with an external diameter of 60 ±1 mm, a wall thickness of 5 ± 1 mm and a length of 500 ± 5 mm. If the test substance may react with the steel, the inside of the tube may be coated with fluorocarbon resin. The bottom of the tube is closed with two layers of 0.08 mm thick polyethylene sheet held in place with rubber bands and insulating tape. For samples, which affect polyethylene, polytetrafluoroethylene sheet can be used. The booster charge is a 200 g RDX/wax mixture (95/5), 60 ± 1 mm in diameter and about 45 mm long with a density of 1600 ± 50 kg/m^3. Additional information on the explosive behaviour of the test sample can be gained by the use of a witness plate, as shown in Figure 8. The mild steel witness plate, 150 mm square and 3.2 mm thick, is mounted at the upper end of the tube and separated from it by spacers 1.6 mm thick.

The sample is loaded to the top of the steel tube. Solid samples are loaded to the density attained by tapping the tube until further settling becomes imperceptible. The sample mass is determined and, if solid, the apparent density calculated. The density should be as close as possible to the shipping density. The tube is placed in a vertical position and the booster charge is placed in direct contact with the sheet, which seals the bottom of the tube. The detonator is fixed in place against the booster charge and initiated. Two tests should be performed unless detonation of the substance is observed.

The test results are assessed on the basis of the type of fragmentation of the tube; and if the occasion arises, the measured rate of propagation in the substance.

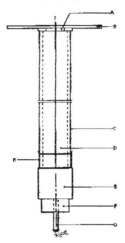

(A) Spacers (B) Witness plate
(C) Steel tube (D) Substance under test
(E) RDX/wax or PE1TI/TNT booster charge (F) Detonator holder
(G) Detonator (H) Plastics sheet
(J) Velocity probe

Figure 8: UN DETONATION TEST

The experiments showed a decaying detonation wave. Initially, a shock waves velocity of 3.2 km/s in 4 the first experiment and 3.8 km/s in the second experiment was found. At a distance of about 350 mm from the booster, the velocity rapidly decreased. "The fragmented length of the tube was about 400 mm. The recovered fragments were characteristic of a very violent deflagration. From the test results it was concluded that the slurry is not capable of propagating a detonation.

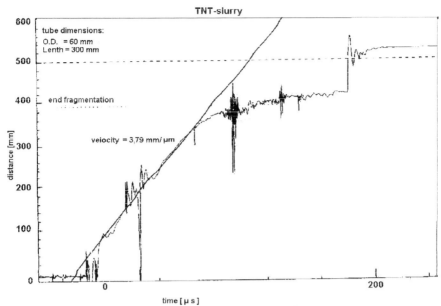
Figure. 9 Detonation test with TNT slurry

3.5. DDT TEST

This test is used to determine the tendency for the substance to undergo transition from deflagration to detonation.

The apparatus consists of a seamless steel tube (A37 type), id. 40.2 mm, wall thickness 4.05 mm, and length 1200 mm. The static resistance of the tube is 74.5 MPa. As shown in Figure 8, two screwed caps close the tube and a probe for monitoring the shock wave velocity is fitted. The tube is placed horizontally onto a lead witness plate of thickness 30 mm. The substance is ignited by a heated wire, composed of Ni/Cr (80/20) with diameter 0.4 mm and length 15 mm, located at one end of the tube.

The test substance is filled into the tube and compacted by hand compression. A current of up to 8 A is used for a maximum of three minutes to heat the ignition wire and ignite the substance. The test is performed three times unless deflagration to detonation transition occurs as shown by compression of the lead witness plate or by the measured propagation velocity.

The test result is considered "positive" if in any trial detonation occurs. Evidence of detonation may be assessed by whether the lead witness plate is compressed in a manner characteristic of detonation; and the measured propagation velocity is greater than the speed of sound in the substance and constant in the part of tube furthest from the initiator.

The length before detonation and the detonation velocity should be noted. The test result is considered "negative" if the witness plate is not compressed and the speed of propagation, if measured, is less than the speed of sound in the substance.

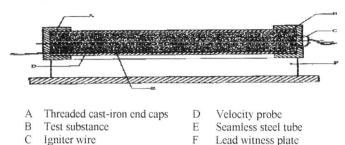

A Threaded cast-iron end caps   D Velocity probe
B Test substance   E Seamless steel tube
C Igniter wire   F Lead witness plate

Figure 10: DDT TEST

The density of the slurry in the test was 1200 kg/m; An amount of 5 gram of black powder was used to ignite the slurry. After ignition of the black powder, no external changes to the tube were visible. After ample time, one screw cap was opened. It was noted that there was an internal pressure in the tube, probably as a result of the burning of black powder. The TNT slurry was blackened over a length of more than 100 mm. From these observations it was concluded that no DDT had occurred.

3.6. KOENEN TEST

The apparatus consists of a non-reusable steel tube, with its reusable closing device, installed in a heating and protective device. The tube is deep drawn from sheet steel of suitable quality. The mass of the tube is 25.5 ±1.0 g. The dimensions are given in Figure 11. The open end of the tube is flanged. The closing plate with an orifice, through which the gases from the decomposition of the test substance escape, is made from heat-resisting chrome steel and is available with the following diameter holes: 1.0 - 1.5 mm, 2.0 mm, 2.5 mm, 3.0 mm, 5.0 mm, 8.0 mm, 12.0 mm and 20.0 mm. The dimensions of the threaded collar and the nut (closing device) are given in Figure 12.

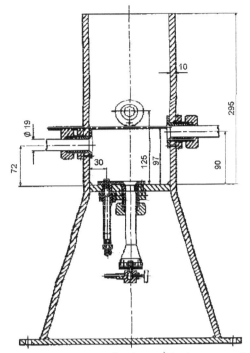

Figure 11 Koenen Test Tube Assembly (measures in mm)

Heating is provided by propane, from an industrial cylinder fitted with a pressure regulator, via a flow meter and distributed by a manifold to the four burners. Other fuel gases may be used providing the specified heating rate is obtained. The gas pressure is regulated to give a heating rate of 3.3 ± 0.3 K/s when measured by the calibration procedure. Calibration involves heating a tube (fitted with a 1.5 mm orifice plate) filled with 27 cm^3 of dibutyl phthalate. The time taken for the temperature of the liquid (measured with a 1 mm diameter thermocouple centrally placed 43 mm below the rim of the tube) to rise from 50°C to 250°C is recorded and the heating rate calculated.

Because the tube is likely to be destroyed in the test, heating is undertaken in a protective welded box, the construction and dimensions of which are given in Figure 10. The tube is suspended between two rods placed through holes drilled in opposite walls of the box. The arrangement of the burners is given in Figure 10. The burners are lit simultaneously by a pilot flame or an electrical ignition device. The test apparatus is placed in a protective area. Measures should be taken to ensure that any draughts does not affect the burner flames. Provision should be made for extracting any gases or smoke resulting from the test.

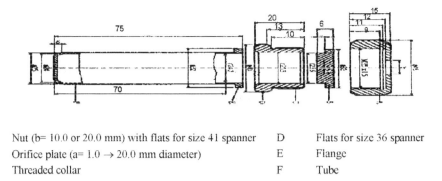

A	Nut (b= 10.0 or 20.0 mm) with flats for size 41 spanner	D	Flats for size 36 spanner
B	Orifice plate (a= 1.0 → 20.0 mm diameter)	E	Flange
C	Threaded collar	F	Tube

Figure 12. Koenen Heating and Protective Device (measures in mm)

Normally substances are tested as received, although in certain cases it may be necessary to test the substance after crushing it. For solids, the mass of material to be used in each test is determined using a two-stage dry run procedure. A tarred tube is filled with 9 cm^3 of substance and the substance tamped with 80 N force applied to the total cross-section of the tube. If the material is compressible then more is added and tamped until the tube is filled to 55 mm from the top. The total mass used to fill the tube to the 55 mm level is determined and two further increments, each tamped with 80N force, are added. Material is either added, with tamping, or taken out as required to leave the tube filled to a level 15 mm from the top.

A second dry run is performed, starting with a tamped increment of one third of the total mass found in the first dry run. Two more of these increments are added with 80 N tamping and the level of the substance in the tube adjusted to 15 mm from the top by addition or subtraction of material as required. The amount of solid determined in the second dry run is used for each trial filling being performed in three equal increments, each compressed to 9 cm^3. (This may be facilitated by the use of spacing rings.) Liquids and gels are loaded into the tube to a height of 60 mm taking particular care with gels to prevent the formation of voids. The threaded collar is slipped onto the tube from below, the appropriate orifice plate is inserted and the nut tightened by hand after applying some molybdenum disulphide based lubricant. It is essential to check that none of the substance is trapped between the flange and the plate, or in the threads.

With orifice plates from 1.0 mm to 8.0 mm diameter, nuts with an orifice of 10.0 mm diameter should be used; if the diameter of the orifice is above 8.0 mm, that of the nut should be 20.0 mm. Each tube is used for one trial only. The orifice plates, threaded collars and nuts may be used again provided they are undamaged.

The tube is placed in a rigidly mounted vice and the nut tightened with a spanner. The tube is then suspended between the two rods in the protective box. The test area is vacated, the gas supply turned on and the burners lit. The tune to reaction and duration of reaction can provide additional information useful in interpreting the results. If rupture of the tube does not occur, heating is to be continued for at least five minutes before the trial is finished. After each trial the fragments of the tube, if any, should be collected and weighed.

The following effects are differentiated:

"O": Tube unchanged;

"A": Bottom of tube bulged out.

"B": Bottom and wall of the tube bulged out

"C": Bottom of the tube split.

"D": Wall of the tube split.

"E": Tube split into two fragments.

"F": Tube fragmented into three or more mainly small pieces, which in some cases may be connected with each other by a narrow strip.

"G": Tube fragmented into many mainly small pieces, closing device undamaged, and

"H": Tube fragmented into many very small pieces, closing device bulged out or fragmented.

The series of trials is started with a single trial using an orifice plate of 20.0 mm. If, in this trial, the result "explosion" is observed, the series is continued with trials using tubes without orifice plates and nuts but with threaded collars (orifice 24.0 mm). If at 20.0 mm "no explosion" occurs, the series is continued with single trials using plates with the following orifices 12.0 mm, 8.0 mm, 5.0 mm, 3.0 mm, 2.0 mm, 1.5 mm and finally 1.0 mm until, at one of these diameters, the result "explosion" is obtained. Subsequently, trials are carried out at increasing diameters, according to the sequence given in 3.5.1, until only negative results in three tests at the same level are obtained. The limiting diameter of a substance is the largest diameter of the orifice at which the result "explosion" is obtained. If no "explosion" is obtained with a diameter of 1.0 mm, the limiting diameter is recorded as being less than 1.0 mm.

The result is considered "+" if the substance shows some reaction on heating under confinement if the limiting diameter is 1.0 or more. The result is considered "-" if the substance shows no reaction on heating under confinement if the limiting diameter is less than 1.0 mm.

With the TNT slurry, the limiting diameter was less than 1.0 mm. No fragmentation of the tubes occurred.

3.7. TPT

The time/pressure apparatus (Figure 13) consists of a cylindrical steel pressure vessel 89 mm in length and 60 mm in external diameter. Two flats are machined on opposite sides (reducing the cross-section of the vessel to 50 mm) to facilitate holding whilst fitting the firing plug and vent plug. The vessel, which has a bore of 20 mm diameter, is internally rebated at either end to a depth of 19 mm and threaded to accept 1" British Standard Pipe (BSP). A pressure take-off, in the form of a side arm, is screwed into the curved face of the pressure vessel 35 mm from one end and at 90° to the machined flats. The socket for this is bored to a depth of 12 mm and threaded to accept the 1/2" BSP thread on the end of the side arm. A washer is fitted to ensure a gas tight seal. The side arm extends 55 mm beyond the pressure vessel body and has a bore of 6 mm. The end of the side arm is rebated and threaded to accept a diaphragm type pressure transducer. Any pressure device may be used provided that it is not affected by the hot gases or decomposition products and is capable of responding to rates of rise of 690-2070 kPa in not more than 5 ms.

The end of the pressure vessel furthest from the side-arm is closed with a firing plug which is fitted with two electrodes, one insulated from, and the other earthen to, the plug body. The other end of the pressure vessel is closed by aluminium bursting disk 0.2 mm thick (bursting pressure approximately 2200 kPa) held in place with a retaining plug, which has a 20 mm bore. A soft lead washer is used with both plugs to ensure a good seal. A support stand holds the assembly in the correct attitude during use. This comprises a mild steel base plate measuring 235 mm x 184 mm x 6 mm and a 185 mm length of square hollow section (S.H.S.) 70 x 70 x 4 mm.

The ignition system consists of an ignition wire (Nichrome), together with a 13 mm square piece of primed cambric. Primed cambric consists of a linen fabric coated on both sides with a potassium nitrate/silicon/sulphurless gunpowder pyrotechnic composition.

The apparatus, assembled complete with pressure transducer but without the aluminium bursting disk in position, is supported firing plug end down. 5.0 g of the substance is introduced into the apparatus so as to be in contact with the ignition system. Normally no tamping is carried out when filling the apparatus unless it is necessary to use light tamping in order to get the 5.0 g charge into the vessel. Note should be taken of the charge weight used. The lead washer and aluminium bursting disk are placed in position and the retaining plug is screwed in tightly. The charged vessel is transferred to the firing support stand, bursting disk uppermost, which should be contained in a suitable, armoured fume cupboard or firing cell. A voltage source is connected to the external terminals of the firing plug and the charge is fired. The signal produced by the pressure transducer is recorded on a suitable system, which allows both evaluation, and a permanent record of the time/pressure profile to be achieved (e.g. transient recorder coupled to a chart-recorder).

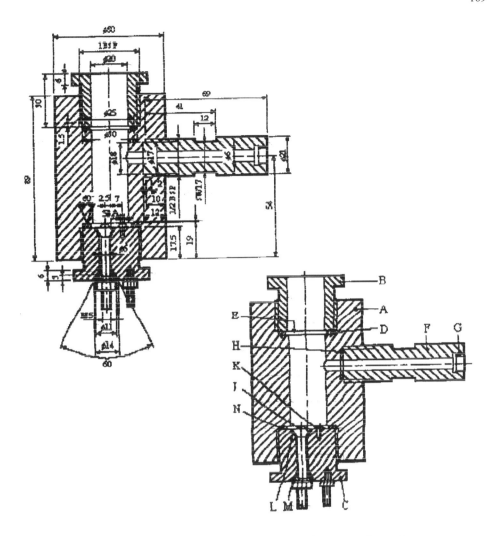

Figure 13 Time Pressure Test apparatus (measures in mm)

A	Pressure vessel body	H	Copper washer
B	Bursting disk retaining plug	J	Insulated electrode
C	Firing plug	K	Earthed electrode
D	Soft lead washer	L	Insulation
E	Bursting disk	M	Steel cone
F	Side arm	N	Washer distorting groove
G	Pressure transducer thread		

The test is carried out three times. The time taken for the pressure to rise from 690 kPa to 2070 kPa above atmospheric is noted. The shortest time interval should be used for classification.

The test results are interpreted in terms of whether a gauge pressure of 2070 kPa is reached and if that is the case, the time taken for the pressure to rise from 690 kPa to 2070 kPa gauge is measured.

The result is considered "+" and the substance to show the ability to deflagrate if the maximum pressure reached is greater than or equal to 2070 kPa. The result is considered "-" and the substance to show no likelihood of deflagration if the maximum pressure reached in any one test is less than 2070 kPa gauge.

The test with the TNT slurry was performed three times. In all three cases the pressure of 2070 kPa was not reached. It was concluded that the effect of ignition under confinement is only very minor.

3.8. CAP SENSITIVITY

The experimental set up for the cap sensitivity test is shown in Figure 14 and consists of a cardboard tube of minimum diameter 80 mm and length 160 mm with a maximum wall thickness of 1.5 mm, closed at the base with a membrane just sufficient to retain the sample. The intense mechanical stimulus is provided by a standard detonator inserted centrally in the top of the explosive in the tube to a depth equal to its length. Below the tube is the witness, which consists of a 1.0 mm thick 160 x 160 mm steel plate, placed on a steel ring of 50 mm height, 100 mm inner diameter and 3.5 mm wall thickness. The apparatus is placed onto a square shaped steel plate of 25 mm thickness and 152 mm sides.

The substance under test is filled into the tube in three equal increments. For free-flowing granular substances, the sample is consolidated by allowing the tube to fall vertically through a height of 50 mm after filling each increment. Gel-type substances are carefully packed to prevent adding voids. In all cases, the final density of the explosive in the tube should be as close as possible to its shipping density. For high-density cartridged explosives with a diameter greater than 80 mm, the original cartridge is used. Where such original cartridges are inconveniently large for testing, a portion of the cartridge not less than 160 mm long may be cut off and used for testing. In such cases the detonator is inserted into the end in which the substance has not been disturbed by the action of cutting the cartridge. The tube is placed onto the witness and steel base plate and the standard detonator inserted centrally into the top of the explosive. The detonator is then fired from a safe position and the witness examined.

The test with the TNT slurry was conducted two times; no indentation of the witness plate was visible. It was concluded that the slurry is not sensitive to intense mechanical stimuli.

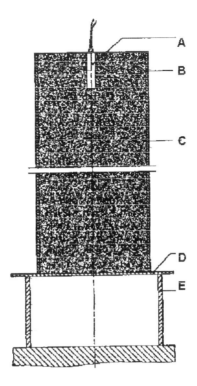

Figure 14. Cap sensitivity test

A	Detonator	D	Witness late
B	Tube fireboard	E	Steel ring
C	Test substance		

3.9. SENSITIVITY TO ELECTROSTATIC DISCHARGE

The sensitivity of a substance to spark is determined by the reaction of that substance towards a capacitor discharge with certain energy. The apparatus is built according to the design of the Explosives Research Laboratory of the Bureau of Mines, Pittsburgh, USA [4]. It consists of a brass cylinder in which a capacitor, a resistance (optional) and a needle are positioned co-axially in series. Underneath this, separated with a space of 3 mm, a circular ground plate is located. The ground plate is connected to earth through a vacuum switch. The sample container is placed under the needle, after which the capacitor is charged through a resistance of 10 MΩ to the desired voltage. After closing the vacuum switch a spark will jump over from the needle through the sample to the ground plate, provided the energy is sufficiently high. Three discrete energy levels can be obtained by this particular version of the apparatus: 0.045 J; 0.45 J and 4.5 J. The reaction of the substance is observed, together with indications of local reactions like colour change, melting, hardening and decomposition.

The slurry, which contains water, was found to be not sensitive to sparks. In none of the tests a reaction could be noted. The dried slurry was a little more sensitive to sparks, i.e. with spark energy of 4.5 J a reaction occurred.

4. Discussion and conclusion

The sensitivity of the slurry and even the dried slurry towards impact and friction is low, the sensitivity towards electrostatic discharge of the slurry is low as well. The dried slurry is a little more sensitive for Electro Static Discharge but still has an acceptable level. This means that the risk to ignite the slurry during the handling is insignificant. Even the dried slurry that can be formed in the process due to accidental spilling of the slurry is hard to ignite.

Heating and ignition under confinement gives very favourable results, i.e. no deflagration of the slurry occurs. This means that if during the handling of the slurry in the process equipment an external heating source occurs, which may be caused due to accidental fire in the building, short-circuiting of the electrical wiring from the process equipment or the ignition of dried slurry outside the equipment, no propagating reaction in the slurry present in the equipment can be expected.

The test results further showed that no transition from deflagration to detonation occurred. This means that even under strong confinement from the process equipment, a starting decomposition caused by external fire cannot grow into a detonation.

The slurry is not capable of detonation. Only in the case of an intense shock wave striking the slurry, a violent deflagration might occur. Such an intense shock wave can only be generated by the detonation of another explosive. In fact this is the scenario of a terrorist attack.

Based on all test results it can be concluded that the slurry mixture can safely be handled and processed in the process equipment (Fluidised Bed Oven and related equipment). Furthermore the slurry can be transported and stored as a non-explosive, under the regime of hazardous chemicals.

References

1. TNO Patent 1007664
2. United Nations, New York and Geneva, (1995), Recommendations *on the Transport of Dangerous Goods, Manual of Tests and Criteria,* Second Revised Edition. (ST / SG / AC.10 / 11 / rev2)
3. Stanag 4479, (1995), Explosives Vacuum Stability Test, Edition 1, Oct. 1995.
4. Brown, Ed. F.W., Kusler, D.J., and Gibson, F.C. (1953), Sensitivity of Explosives to Initiation by Electro Static Discharges, *Bureau of Mines Report of Investigations 5002* Sept. 1953, USA.
5. TNO Report 1999-IN 14, Optimisation of Incineration of Explosives in a Fluidized Bed Oven.

USE OF CONVERTED HIGH ENERGY VALUE EXPLOSIVE MATERIALS AS INDUSTRIAL ENERGETIC MATERIALS

N.K.SALYGIN, B.V.MATSEEVICH, V.P.GLINSKI, O.F.MARDASOV, N.I.PLECHANOV
FGUP "KNIIM". Krasnoarmeisk, Russia

The unitary approach to the problem of utilisation of ammunition determined the task of overworking high-energy value explosive materials as trinitrotoluene-hexogene mixtures, trinitrotoluene-hexogene-aluminium mixtures and phlegmatised hexogene-aluminium. One of basic conditions of the use of utilised explosives as industrial explosives is the handling safety. Quantitative characteristics of this feature could be evaluated by use of the sensitiveness to friction and impact. Some values of the sensitiveness to friction and impact for certain explosive materials are given in Tab. 1.

TABLE 1. Sensitiveness to friction and impact

	Explosive material	Sensitiveness to friction, MPa	Sensitiveness to impact on apparatus No. 1
1	Mixtures of trinitrotoluene-hexogene (TH)	500	48
2	Mixtures of trinitrotoluene-hexogene-aluminium(THA)	400	40
3	Phlegmatised mixtures of hexogene-aluminium	400	76
4	Gunpowder	100 – 150	50 – 60
5	Trinitrotoluene (TNT)	>500	4 - 8

Here trinitrotoluene was selected as a standard measure for the handling safety. From this data follows, that according to the sensitiveness to friction and impact, the high-energy value explosives may not be used as industrial explosives without additional modification. The simplest solving of the reduction of sensitiveness to mechanical actuation is the phlegmatisation.

This problem was investigated in the Institute KNIIM together with the study of methods and technology of extraction of explosive materials from cartridges, i.e. the phlegmatisation was usually carried out in the course of the ammunition disassembly together with additional explosive material overwork and modification to industrial energetic material.

The diagram of the process of industrial explosives making is shown in Figures 1 and 2. Characteristic data are given in Table 2.

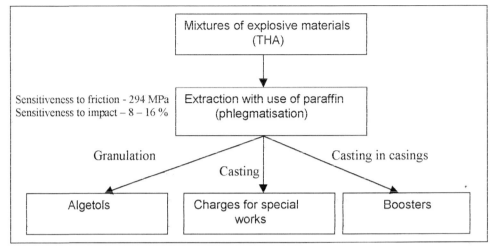

Figure 1

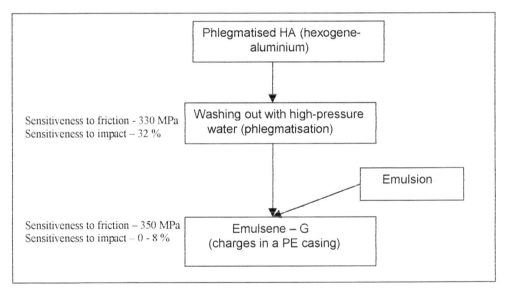

Figure 2

From the Table 2 results, that in accordance with the sensitiveness to friction and impact, industrial explosives manufactured on the basis of utilised high-energy value

Explosive material	Characteristics						
	Explosion temperature kJ/kg	Flash point, °C	Sensitiveness to impact, according to GOST 4545, on the apparatus No.1	Sensitiveness to friction, OST 84-895, on the apparatus No.1	Density, kg/m³	Detonation velocity km/s	Nspec, (kW/m²) 10⁹
Algeton – 15	4735	210	8	294	900 – 1000 (bulk)	4,6	21,78
Algeton – 25	4860	210	8	294	900 – 1000 (bulk)	4,8	23,33
Algeton – 35	4986	210	16	294	900 – 1000 (bulk)	5	24,93
Emulsen G	4921	230 – 240	0 – 8	350	1450 – 1480	5,4 – 6,0	36,08
Trinitrotolu ene	3900	295 – 305	4 – 8	>500	750 – 800(bulk)	5,0 – 5,5 (filled with water)	17,16

Table 2. Characteristics of industrial energetic materials

explosive materials are at the same level, as trinitrotoluene. However, with their energetic characteristics they exceed trinitrotoluene. In present time a new technology, ND and equipment for algetols was developed and elaborated and a production unit at FGUP "KNIIM" and DVPO "Voschod" was erected. Algetols are supplied by "Transvzryvprom" Co, where they are used as borehole charges. For emulsene G a new ND was elaborated too. General view of the charge is shown in Figure 3,

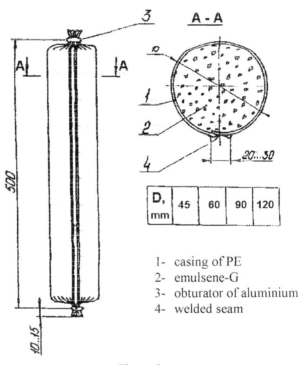

1- casing of PE
2- emulsene-G
3- obturator of aluminium
4- welded seam

Figure 3

The overwork of gunpowder to industrial explosive was realised in Russia usually by the way of phlegmatisation by rock-oil products or water-salt solutions (granipor, zernite, dibasite etc.). Such energetic materials are used in the form of borehole charges. Before phlegmatisation the tube powder should be pulverised. Equally with traditional techniques proposed KNIIM the use of gunpowder as sensibilization component for emulsion industrial explosives or for casting compounds type porotol – mixtures of trinitrotoluene (melt) and granular pyroxilin powder in the ratio of 50:50 and product manufactured of them: boosters (DPU), seismic survey charges etc. General scheme of this process is shown in Figure 4, the section of the charge in Figure 5.

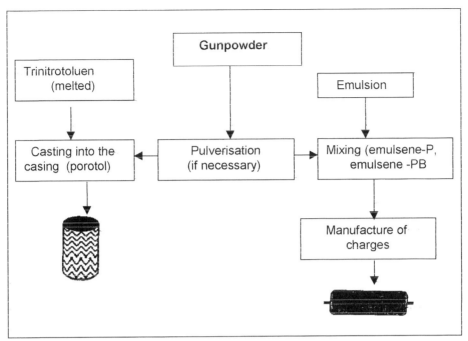

Figure 4

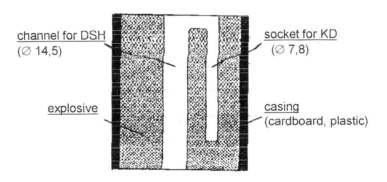

Figure 5

Boosters (DPU) are produced with a mass of 700, 800, 900 and 1000 g.

Physico-chemical and detonation characteristic of poroton.

- Specific detonation heat, kJ/kg 3875
- Detonation velocity, m/s................................6500

- Specific volume of gases, l/kg 812
- Disruptive force by GOST 5984 30,8
- Detonation pressure, Gpa 16-20
- Density, kg/m^3 1500
- Specific power, kW/m^2 .... 37,7 x 10^9

Charges of porotol can be reliably initialised by KD No.8, DSH, by non-electric systems of initiating type "Nonel" after exposition in water under pressure of 0,3 MPa in the course of 15 days. The reliable initiate borehole charges of water containing explosives at the temperature of 85°C and even the charges of explosives, containing diesel fuel (ANFO).

THE TECHNOLOGY OF SLURRY AS DISPERSING MEDIA FOR POWDERS AND PROPELLANTS

J.C.LIBOUTON
Centre de Recherches, Nobel Explosifs Belgique

ABSTRACT

In the field of civilian explosives slurries have to deliver maximum power density.
The highest density, one can use to achieve a cap sensitivity, assesses the good choice of the formulation. Four ways can explain the sensitising of the slurry by passing:
- Lower sound velocity
- Hot spot formation
- Adiabatic compression
- Differential flow velocity

Addition of dry materials doesn't desensitise the slurry and if the particle is heavy, one can even enhance the sensitivity.
The target is then to define a good matrix with the nominal crystallisation.
Thermodynamic choice of the fuel can be done on the ratio between hat of combustion and oxygen balance. Formulations with propylene glycol and calcium nitrate ease to get a cap sensitivity at density as high as 1.2. Other ways to lower the fudge point of the solution are addition of urea to formaldehyde.
With such mixtures, additions as high as 50 % of dry materials can be done.

1. Explosives based on ammonium nitrate

For civilian explosives whose utilisation is done in borehole, the measure of efficiency is the power density. This power density is a function of the section of the cartridge, the density of the explosive, the detonation velocity and the heat release of the formulation:

$$Powerdensity = Section(m^2)*Density(kg/m^3)-Detonationvelocity(m/s)*Q_v(J/kg)$$

The parameters on which we have to work are the density, the detonation velocity and the energy.

1.1 DENSITY

Usually, explosives based on ammonium nitrate have a dependence of the detonation velocity versus density for a given diameter in form of parabola whose the maximum is figure of merit of the formulation. Really, the density is not true parameter, the true parameter is the porosity but generally speaking density is a good indicator.

Higher will be the density at which one measure the maximum velocity for given initiation, better will be the formulation.

Higher will be the density, better will be the sensitivity of the formulation. This sensitivity can be due to the intimacy of the contact between fuel and oxidiser size of oxidiser particle) or to the kinetic of fuel oxidation in the chain reaction (use of partially oxidised fuel as alcohol) (increase of atomic hydrogen in the pool of radicals).

1.2 DETONATION VELOCITY

Detonation velocity of the mixture can be computed by modelisation. Use of this kind of program shows a dependence of detonation velocity versus the gas volume evolved by the formulation. Since water has among usual gaseous molecules the lowest molecular weight (hydrogen excluded), water involved in ammonium nitrate slurry can enhance the detonation velocity.

1.3 ENERGY

Addition of water to the formulation decreases the energy. However, addition of aluminium can increase the energy, the temperature of the gases and then the expansion work. A drawback is the diminution of the gasses volume.

Another way to increase the energy is addition of dry materials including propellants or explosives wastes to the formulation.

Advantages of ammonium nitrate based explosives are the low shock and friction sensitivity. Moreover, generally the thermal stability is very good.

The difficulty is to achieve a good shelf life and good cap sensitivity.

2. History of the development of ammonium nitrate based explosives

One can summarise four steps during evolution of these products:
- In a first generation, to achieve a minimum sensitivity one added explosive components to reticulated slurry of ammonium nitrate. Usually, the sensitivity was not very good (no cap sensitive) but the detonation velocity and the density were high.
  Addition of miscible methyl amine nitrate to the slurry cased Dupont to achieve cap sensitivity at density of 1.2.

Characteristics of methyl amine nitrate is lower the fudge point of the solution and then to decrease the amount of crystallisation in the slurry.

- In a second generation, under the impulsion of Dr Cook from Ireco, it appeared that addition of painting grade aluminium achieved the cap sensitivity. The bubble of gas catched on the surface of aluminium particle sensitised the slurry up to density of 1.2. Cumulative effect of temperature generated by the shock compression of the bubble, form of the particle (foil of low thickness with rupture of the foil during oxidation) and high-energy release of aluminium by low oxygen consumption. Drawback of these formulations were: hydrostatic desensitising for high depth of underwater loading, dynamic desensitising by compression coming from others mines of the some shot and channel effect in underground.

- The third generation was mainly a research work The target was to suppress addition of explosives or aluminium to achieve a cap sensitivity. Explosives coming from this research were few used because the discovery of emulsion make these products obsolete. The goal was to lower the fudge point by addition of urea, formaldehyde or crystal modifiers as aliphatic amine.

- The fourth generation is emulsion, water in oil emulsion cases to suppress the formation of crystals during cooling and then to enhance the contact between oxidiser and fuel. The cap sensitivity of emulsion can be achieved by porosity inside the matrix.

3. Sensitising by gassing

3.1. POROSITY VERSUS DENSITY

The main factor of sensitising is not the density of the explosive but the porosity inside and the dispersion of this porosity.
One example of this effect is the addition of iron powder to an emulsion matrix. Trials were done unconfined in diameter of 30 mm.
Explosive is initiated by a detonator No 8.
For this type of emulsion, one needs 2 % of microballoons to sensitise the formulation, with an addition of 20 % iron one needs only 1.5 %.
An increase of density by iron addition doesn't desensitise the emulsion if sufficient amount of void allows a pdv mechanical work on the emulsion.

It comes out of these trials (see Table 1.) that an emulsion can be cap sensitive at density as high as 1,23 if a sufficient amount of MB is added.

TABLE 1

	0 %MB	1 %MB	1,5 %MB	1,75 %MB	3,5 %MB
Iron percentage			20	20	20
Emulsion percentage			78,5	78,25	76,5
Energy, Qv (cal/g)			713,5	713,7	699
Experimental density			1,49	1,40	1,17
Theoretical density			1,5	1,46	1,23
Detonation velocity (m/s)			3788	4237	3846
Iron percentage	30	30	28,5	30	30
Emulsion percentage	70	69	70	68,25	66,5
Energy, Q_v (cal/g)	661	655	651	650	638
Experimental density	1,91	1,7	1,6	1,61	1,21
Theoretical density	1,98	1,73	1,61	1,58	1,32
Detonation velocity (m/s)	-	-	4098	3968	3333
Iron percentage				40	
Emulsion percentage				58,25	
Energy, Qv (cal/g)				582	
Experimental density				1,73	
Theoretical density				1,76	
Detonation velocity (m/s)				3731	

3.2 HOT SPOT FORMATION

Let suppose a hot plate of 4000°C (T) in contact with an emulsion at 20°C (To). What could be the time needed to increase the temperature of the emulsion by 100°C (θ) to a distance of 10 μ.

$$\theta = T - (T - T_0)\phi(\frac{x}{2\sqrt{at}})$$

$$\frac{\delta^2\theta}{\delta x^2} = \frac{1}{a}\frac{\delta\theta}{\delta t}$$

where Φ is error function

$$a = \lambda/\rho c$$

p(initial density) = 1,54 g/cm^3
c(specific heat) = 0,613 cal/g/°C
λ(thermal conductivity) = 1,3.10^3 cal/s.cm.°C
x - 10 μ
I = 20 °C, T$_0$ = 4000 °C
Θ = 72,3 μs

Such high value of time for a so low increase of temperature involves others factors (friction streamlines) for ensure a homogeneous temperature.

This theory has been described by Erkman. This computation needs three parameters, the bulk sound velocity (C_0), and the S coefficient in the shock polar and the Gruneisen coefficient [1].

$$U_s = C_0 + S.U_p$$

$$P = \frac{C_0^2.(V_0-V)}{(V_0+S.(V-V_0))^2}$$

$$P_{por} - P_{comp} = \Gamma \frac{1}{V}(E_{por} - E_{comp})$$

$$P_{por} = P_{comp}.[\frac{1-\Gamma/2V.(V_{0comp}-V)}{1-\Gamma/2V.(V_{0por}-V)}]$$

With this approach one can compute the temperature increase from the mechanical work: $P_{por} * \Delta V$.

For a same emulsion of different initial density one can compute the temperature increase behind the shock.

TABLE 2

Density (g/cm3)	VOD (m/s)	Up (m/s)	Us-Up (m/s)	Δ T (°C)
0,73	2923	1730	1193	624
0,78	3280	1889	1391	735
0,88	3731	2028	1703	834
0,94	3970	2080	1890	872
1,01	4274	2147	2127	922
1,04	4459	2206	2253	971
1,1	4661	2216	2445	976

Two opposite factors influence the temperature increase; a low initial density increases the mechanical work by variation of volume but decreases the detonation velocity and the pressure induced. This level of temperature can explain by chain reaction the observed induction time and reaction length.

4. Choice of the fuel

Ammonium nitrate based explosive holds mainly ammonium nitrate, complement is a combustible. Heat release results mainly from the combustible. More one adds combustible in keeping the good stoichiometry, more one pull out heat release.

Comparison between different combustibles includes surprising comments, if one divides the combustion energy (Q_v) by the oxygen balance of the combustible (mass of oxygen needed to burn 100 g), one pick out an index whose the value is roughly the same for different combustibles. This index increases few for alcohol, strongly for nitrated molecule and very strongly for metals,

Table 3 shows the value of this index for different molecules.

5. Applied formulation on basis of calcium nitrate and propylene glycol

Use of partially oxidised molecule, as propylene glycol is good kinetically due to the easier ablation of atomic hydrogen versus aliphatic molecules. Thermodynamically this product is as good as aliphatic.

In the case of slurry where the crystallisation works mainly, use of calcium nitrate inhibits the typical formulation of needles crystal of ammonium nitrate.

Difficulties arise for the thickening of the slurry by gums. A good choice of synthetic gum is necessary. Typically slurry based on these two components can be cap sensitive until density of 1,2. Results of measurement of the detonation velocity versus the total density are plotted on the graph in Figure 1.

Solution: 82,135 %
    H_2O: 7,17
    $NaNO_3$: 5
    $Ca(NO_3)_2NH$: 30
    NH_4NO_3: 38
    EG: 1
    HP8 : 0,165
    XB23 : 0,1
    Catalyst : 0,4
Density : 1,58
Fuel phase: 17,0
    PG: 7,6
    Formic acid: 0,3
    Aluminium: 10
Additives:
    Crosslinking agent
    Gassing agent

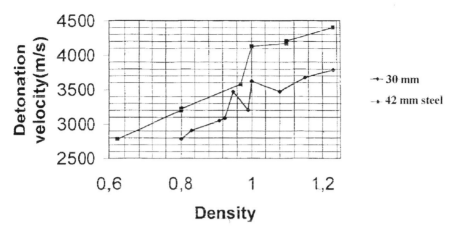

Figure 1 Detonation velocity of a propylene-glycol slurry

Name	Formula C**H**O**N**XX*XX*	Q comb. cal/g	Oxygen Balance	Gold Index
Glycol	C2 H6 O2 N0 **0**0	4115	-128,84	31,93883887
Glycerol	C3 H8 O3 N0 **0**0	3865	-121,57	31,79238299
Starch	C6 H10O5 N0 **0**0	3880	-118,37	32,77857565
Solid paraffin	C24H50O0 N0 **0**0	10520	-344,78	30,51221069
Liquid paraffin	C10H22O0 N0 **0**0	10600	-348,48	30,41781451
Wood meal	C15H22O10N0 **0**0	4390	-136,97	32,05081405
Castor oil	C18H34O3 N0 **0**0	9060	-267,93	33,81480237
Pentaérythritol	C5 H12O4 N0 **0**0	3915	-140,97	27,77186635
Urea	C1 H4 O1 N2 **0**0	2497	-79,89	31,25547628
Hexaméthylentétramin	C6 H12O0 N4 **0**0	6742	-205,36	32,83015193
Gilsonite	C2 H3 O0 N0 **0**0	11000	-325,8	33,76304481
Diméthyldiphénylurea	C18H16O1 N2 **0**0	7751	-286,2	27,08245982
Diphénylamine	C12H11O0 N1 **0**0	8761	-278,83	31,42057885
Carbon	C1 H0 O0 N0 **0**0	7869	-266,57	29,5194508
Phénanthren	C14H10O0 N0 **0**0	9176	-296,15	30,9842985
Anthracen	C14H10O0 N0 **0**0	9192	-296,15	31,03832517
Glycol propylen	C3 H8 O2 N0 **0**0	5201	-168,15	30,93071662
Oleic acid	C18H34O2 N0 **0**0	8813	-288,79	30,51698466
Méthanol	C1 H4 O1 N0 **0**0	4791	-149,75	31,9933222
Formamide	C1 H3 O1 N1 **0**0	2668	-88,77	30,05519883
Formol	C1 H2 O1 N0 **0**0	3738	-106,53	35,08870741
Nitrométhane	C1 H3 O2 N1 **0**0	2540	-39,3	64,63104326
Alkaterge	C22H41O3 N1 **0**0	8394	-267,6	31,367713
Formic acid	C1 H2 O2 N0 **0**0	1113	-34,75	32,02877698
Ammonium chloride	C0 H4 O0 N1 Cl1**0	933	-44,85	20,80267559
Calcium stearate	C35H70O4 N0 Ca1**0	8223	-274,03	30,00766339
Stearic acid	C18H36O2 N0 **0**0	8833	-292,36	30,2127514
SMO	C24H46O7 N0 **0**0	7028	-229,19	30,66451416
Toluen	C7 H8 O0 N0 **0**0	9699	-312,47	31,03977982
Dinitrotoluen	C7 H6 O4 N2 **0**0	4517	-114,16	39,567274
Trinitrotoluen	C7 H5 O6 N3 **0**0	3487	-73,94	47,15985935
Xylen	C8 H10O0 N0 **0**0	9767	-316,38	30,87110437
Nitrocellulose	C24H31O38N9 **0**0	2332	-38,71	60,24283131
Iron	C0 H0 O0 N0 Fe1**0	1750	-38,18	45,83551598
Aluminium	C0 H0 O0 N0 Al1**0	7380	-88,92	82,99595142
Zinc	C0 H0 O0 N0 Zn1**0	1270	-24,96	50,88141026
Silicium	C0 H0 O0 N0 Si1**0	7300	-113,9	64,09130817
Sulphur	C0 H0 O0 N0 Su1**0	2194	-98,78	22,21097388
Nitropenta	C5 H8 O12N4 **0**0	1838	-10,12	181,6205534
RDX	C3 H6 O6 N6 **0**0	2151	-21,6	99,58333333
Tetryl	C5 H7 O8 N5 **0**0	2840	-47,34	59,99155049
Picric acid	C6 H3 O7 N3 **0**0	2631	-45,38	57,97708242
Nitroguanidin	C1 H4 O2 N4 **0**0	1893	-30,74	61,58100195
Nitrate de guanidine	C1 H6 O3 N4 **0**0	1492	-26,2	56,94656489
Methylamine nitrate	C1 H6 O3 N2 **0**0	2011	-34	59,14705882
Polystyren	C8 H8 O0 N0 **0**0	9552	-307,16	31,09779919
Expancel	C5 H5 O0 N1 Cl2**0	4468	-127,95	34,91989058
Sugar	C6 H12O6 N0 **0**0	3386	-106,53	31,78447386

PROMISING BOOSTERS FOR BLASTING COMMERCIAL EXPLOSIVES IN BOREHOLES

V.P.ILIYN, A.G.GOROKHOVTSEV, Y.S.KULAKEVITCH,
N.I.RABOTINSKY, C.P. SMIRNOV
GosNII "Kristall", Dzerzhinsk, Nizhny Novgorod Region, Russia

Among the charges used for blasting borehole charges of commercial explosives and in seismic survey (for geophysical operations) TNT pressed boosters of various configuration and weight are the most widely used. The boosters are chosen due to the following factors:
- raw material (flaky TNT) is not expensive
- their production process (pressing with high-output mechanized units) is rather simple
- their rather high safety in transportation, handling, mounting, borehole charging (because of TNT low sensitivity to impact, friction, exposure to heat)
- their rather high power (because of TNT with 1.55 g/cc density applicable energetic characteristics) for reliable initiation of a low sensitive commercial explosive borehole charge.

However, in spite of waterproofing coating, the boosters have a certain disadvantage - low water resistance that is only several days (water resistance here means that the boosters retain their susceptibility to initiation after being in water under specific conditions). This disadvantage can be explained by large water absorption of the boosters because of their porous structure. Our investigations showed that water absorption of commercial waterproofed pressed TNT charges with 1.55-1.56 g/cc density can be of the following values:

- about 3% after 6 days of keeping them in water under a pressure of 2 atm (0.2 MPa)
- about 3.5% after 6 days of keeping in water under a pressure of 5 atm (0.5 MPa)
- - about 4% after 10 days of keeping in water under a pressure of 5 atm (0.5 MPa).

This causes "phlegmatization" of pressed TNT boosters and their possible subsequent failures in blasting of watered boreholes with all negative after-effects. So it seems justified that consumers want to have charges with such susceptibility to initiation that does not depend (or only slightly depends) on their keeping under water. Besides, in connection with demilitarized TNT appeared recently in large amount, topical task now is to achieve its wide use in commercial blasting charges. One of the possible approaches to resolve this problem is to use casting instead of pressing in production of boosters that significantly reduces their porosity and ability to absorb moisture. Thus the ability of our THPh-850 E and THPh-500 E boosters (Fig. 1) to absorb water is not more than 1.5% even under severe conditions of testing (5 atm of water pressure and up to 35 days of keeping them under water). Employing TNT/hexogen mixture as an explosive and versatility of our booster construction allow using the boosters. with all means of initiation permitted in Russia (caps and electric detonators, detonating cords, non-electric systems of initiation - Nonel, SINV, Edilin), to initiate detonation of any commercial explosives (powdered, granulated, emulsions and hot flowing). Detonation velocity of THPh-850 E and THPh-500 E boosters is within 7600-7800 m/s range and depends on their real density (1.62-1.65 g/cc). However indicated above advantages of TNT/hexogen cast boosters are accompanied with such disadvantages as their high price (they are 3-4 times as expensive as TNT boosters) and high sensitivity to mechanical stimuli in comparison with TNT boosters. These disadvantages restrain wide use of THP-850 E and THP-500 E boosters in spite of the fact that they are very promising (their current production is about 400 000 pcs per year).

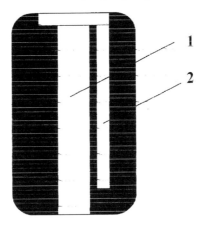

Figure 1. THPh-850 E (THPh-S00 E) booster
1 - channel; 2 - socket

Therefore we thought of more preferable direction for improvement of commercial boosters. It consists of the following: to develop such charges that would retain the advantages of pressed TNT boosters (low cost, safety, large-scale production) and have high reliability in initiation in watered boreholes. This became possible when we developed in GosNII "Kristall" a new process of casting according to which a mould with configuration, corresponding to a booster configuration, was filled with TNT granules and the spaces between them were filled with TNT melted under specific conditions. The charge is produced with almost unlimited water resistance and its strength is more than 2 times as much as that of a charge of pressed TNT flakes. This process can be realized in casting systems with automatic control, the output of which is up to 5-6 million boosters per year and the boosters cost is comparable to that of pressed TNT boosters. T-500 L-KG booster produced using this process (Fig. 2) has 540 g in weight, 62 mm in external diameter and 125 mm in height and is adapted to all permitted in Russia means of initiation.

The investigations showed that after keeping T-500 I,-KG boosters in water for 35 days even under 5 atm pressure their water absorption was not more than 1.3% and they became saturated in 9-10 days.

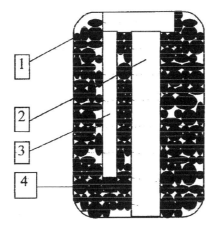

Figure 2. T-S00 L-KG cast booster
1 - TNT granules and melted TNT
2 - channel, 3 - socket, 4 - 1 g of plastic explosive

A wide range of investigations and industrial testing in different geologic and mining conditions indicate high reliability (corresponding to that of THPh-850 E and THPh-500 E hexogen including boosters) in initiating cast T-500 L-KG boosters with different devices that significantly exceeds pressed TNT charges susceptibility to ignition in watered boreholes. Besides T-500 L-KG booster level of safety is considerably higher that that of boosters including hexogen and allows to use procedures of operation adopted for pressed TNT boosters. Detonation velocity of T-500 L-KG boosters is within 6700-6900 m/s range depending on their density.

The process of THPh-850 E, THPh-500 E and T-500 L-KG boosters production includes also the application of demilitarized TNT with solidification point of not lower than that of commercial TNT of A specification (not less than 80°C).

All the boosters developed in GosNII "Kristall" (THPh-850 E, THPh-500 E, T-500 L-KG) are under the protection of RF certificates of authorship and patents and so their production in enterprises other than GosNII "Kristall" is permitted on the base of licensed agreements with GosNII "Kristall". We offer such co-operation both with national and with foreign partners.

THE APPLICATION OF RECLAIMED EXPLOSIVES IN COMMERCIAL EMULSION EXPLOSIVES

N.I. RABOTINSKY, V.A. SOSNIN, V.S. ILIUKHIN
GosNII "Kristall", Dzerzhinsk, Nizhny Novgorod Region, Russia

Ammunition disposal problem is of current concern in connection with conventional armaments reduction and because they have already been stored rather long. The problem can be resolved in three ways: 1) disposal by blasting or incineration, 2) reclamation of explosives and their chemical processing or disposal and 3) reuse of explosives, extracted from ammunition, in commercial explosive compositions. However the analysis of these approaches shows that the first one is economically inefficient and gives environmentally harmful emissions, the second is technically complicated and the third allows reclaiming both explosives and ammunition bodies and so seems rational.

The experience shows that derived wet ingredients of an explosive are better to use in water based explosive compositions. This allows expanding raw material base for production, providing required detonation characteristics and resolving the problem of lowering sensitivity of reclaimed explosive compositions to mechanical stimuli. In Kristall the trend of development water based emulsion explosives prevails so the work was carried out with systems including a matrix and a filling agent with a high explosive. For such systems the following requirements are important:
- chemical compatibility, thermostability and necessary sensitivity to mechanical stimuli;
- optimum ratio of a matrix and a reclaimed explosive for making compositions with performance suitable for various applications in mining operations.

The tests were carried out with model formulations which consisted of an emulsion matrix (the dispersion of water solution of ammonium and calcium nitrates in the mixture of a liquid hydrocarbon and an emulsifier) and a high explosive (pure or reclaimed from ammunition). As a high explosive hexogen (H), phlegmatized hexogen (PH) and a mixture of TNT, hexogen and aluminium (THA) were used. A ratio of the emulsion matrix and the explosive was selected from the range of 80/20 - 50/50. At first the ingredients compatibility was evaluated by heating in a closed volume (Table 1) and then susceptibility to initiation of 36 mm diameter charges in polyethylene film with a charge length to diameter ratio (l/d) of minimum 15 (Table 2).

TABLE 1. Emulsion/explosive mixtures self-ignition delay at heating in ampules

Mixture components, %		Temperature, °C	Time delay of self-ignition, min
Emulsion	Explosive		
100	0	235	235-240
		170	24 hours without self-ignition
0	100PH	170	130-150
50	50PH	170	120 – 130
0	100THA	160	110 – 120
50	50THA	160	50 - 60

TABLE 2. Susceptibility of emulsions with filling agents to initiation

Emulsion mixture composition, %		Detonation velocity(km/s) by initiation with	
Emulsion	Filling agent	Electric detonator No 8	TNT pellet*
100	gas microbubbles	From 3.5 to 4.8	From 3.8 to 4.8
95	5	4.1 – 4.4	4.2 – 4.4
80	50microspheres, 20H	Failure	5.1
70	30H	Failure	5.8
60	30H	Failure	5.8
50	40H	-	6.4
60	40PH	Failure	6.7
50	50PH	Failure	6.5
50	50THA	Failure	5.4
0	100PH	6.4	-

* TNT pressed pellet with 10 g weight and 1.62 g/cc density

The tests showed that compatibility of the emulsion/high explosive mixture ingredients is satisfactory. Lower thermal stability of the composition with THA is clue to the presence of aluminium powder the reaction of which with a dispersion phase of the emulsion seems to initiate decomposition. If there is no heating, the samples can withstand usual time of storage.

High explosives sensitizing ability is deficient for the charges initiated with a commercial electric detonator, i.e. of a cord type. While a high explosive with sufficient initiation impulse gives a positive effect as a sensitizer at relatively low content (20%) and the effect increases with the high explosive content.

The next series of tests was carried out to determine the possibility to increase the emulsion/explosive mixture susceptibility to initiation by addition of microspheres. The results are given in Table 3.

TABLE 3.

Fraction of emulsion total weight, %		Detonation velocity (km/s) by initiation with	
hexogen fraction	microspheres fraction	electric detonator No 8	TNT pellet
15	2	Failure	5.0
20	3	Failure	4.8
25	3	Failure	5.0
20	5	4.6	4.7

The tests showed that the emulsion mixture sensibilization with both a high explosive and microspheres did not give advantage in initiation with an electric detonator.

The same result was with combined sensibilization of the emulsion mixtures with 20% of hexogen preliminary chemically gassed: the matrix with the composition detonation velocity of 4.9-5.1 km/s and the content of gassing solvent from 0.1 to 0.5% and with 1.3-1.5 g/cc density.
Initiation susceptibility of flowing suspension mixtures of slurry type with 20-28% of hexogen was of the same level: they detonated from TNT booster with 4.5-5.3 km/s velocity in charges of 36 mm diameter and 1.4-1.6 g/cc density.

The tests were carried out to evaluate sensitivity to mechanical stimuli of the emulsion/high explosive mixtures. The mixtures were used with the content of 15-50% of hexogen, 50-85% of emulsion, 1.5-3.0% of microspheres or perlit. The results are given in Table 4.

TABLE 4. The emulsion mixtures sensitivity to mechanical stimuli

Emulsion mixture composition, %			Sensitivity to mechanical stimuli		
High explosive	Microspheres	Emulsion	to impact (GOST 4545-88)		to friction
			Lower limit, mm	Frequency of explosions, app.no 1, %	(OST V 84-895-83, gf/cm2, % of explosions
15H	0	85	<200	4	3500/4
15H	3	82	-	4	3000/4
15PH	0	85	250	12	7000/0 8000/4
30PH	0	70	200	40	8000/0 9000/8
50PH	0	50	120	40	4000/0 5000/4
20PH	2 perlit	78	100	36	3000/0 4000/24
20THA	0	80	250	8	6000/0 7000/12
50THA24H	0	50	200	32	6000/0 7000/16
24H	Powder mixture ammonium nitrate and Al	Of TNT	100-200	68	4000/0 5000/0 7000/92
	Ammonite 6ZhV		200	16	7000/4 8000/16 10000/84

The results show, that the mixtures on the base of emulsions filled with high explosives, approximately correspond to Ammonite 6ZhV according to their sensitivity to mechanical stimuli and somewhat lower than Ammonite No 1 for rocks. High explosive fraction of total weight of the mixtures is, within the limits of practical interest, convenient for using in the compositions for mining operations.

Emulsion/reclaimed high explosive mixture compatibility and detonation ability tests showed that it is possible on the base of the mixtures to produce catridged commercial explosives.

The cartridged commercial explosives will have two fields of application: with high power compositions of limited use in charges of 36-45 mm diameter for blasting in mines with severe conditions and as boosters and borehole charges of 60-90 mm diameter in blasting operations on the surface.

TABLE 5. The emulsion mixtures detonation ability

Emulsion mixture composition, %			Characteristics			
Emulsion	Explosive	Air in microspheres, % by vol.	Limited size of explosive particles, mm	Critical diameter of detonation, mm	Detonation velocity (km/s) at density (g/cc)	Detonation transfer, cm
85	15(HPha)	-	3	60	3.8**	-
85	15(HPha)	-	7	60	3.7(1.57-1.55)*	-
85	15(HPha)	25-	3	25	4.4(1.1)	-
75	25(HPha)	-	3	50	4.7(1.48)	3
75	25	-	7	50	4.6(1.45)	3
75	25	25	3	25	4.6(1.2)	5
80	20(H)	-	-	-	5.1	-
75	20(H(	-	-	-	5.3	-
100	-	25	-	25	3.9(1.1)	5

Notes: * Microspheres with air were added in 25% amount in addition to 100% of the emulsion/explosive mixture

** The charge diameter is 80 mm, in the rest of tests it is 60 mm

The results of the tests showed that with the explosive fraction of total weight from 0 to 70% the mixtures are thermostable: the time of heat self-ignition onsets is within the range of 90-235 min for temperatures of 160-235°C for OST 84-1067-81 dry or wet explosives.

The emulsion/explosive mixtures sensitivity to impact and friction is comparative to that of explosives of ammonite 6ZhV type and significantly lower than that of initial reclaimed explosive.

HPha dewatering and mixing with an emulsion matrix was tried out. It was found that the explosive material residual humidity within the range of 14-20% is reached on a vacuum filter for 20 - 30 minutes. Wet composition is successfully admixed to emulsion in a laboratory mixer with blades for 2-5 minutes, if their ratios range from 30/70 to 80/20. There was observed slight separation of water, the amount of which increased with increasing humidity of explosive material and duration of mixing. The separated water influence on the emulsion/explosive mixture characteristics in storage requires carrying out special studies.

Detonation characteristics were tested with the mixtures including 15-35`% of the explosives particle sizes of which were less than 3 mm (fraction 1) and less than 7 mm (fraction 2). Significant influence of particle size on detonation ability was not observed. Critical diameter of detonation for emulsion/explosive mixtures including 15-25% of an explosive was 50-60 mm and when additionally 25% of entrained in microspheres air were introduced it became 25 mm. The detonation velocities of the mixtures were within the range of 3.8-4.7 km/s at the densities of 1.10-1.57 g/cc. Detonation transfer distance for cartridged mixtures through the air is from 0 to 5 cm.

On the base of these results a conclusion is made that it's possible and advantageous to develop compositions for mining operations in two directions:
- articles with 90 mm diameter and more, having 50 mm critical diameter of detonation, insensitive to initiation with an electric detonator and including 15-25% of reclaimed explosive;
- articles with less than 60 mm diameter that are susceptible to initiation with an electric detonator and include 30-35% of recovered explosive and 15-25% by volume of entrained air.

The process of production of the emulsion matrix providing stability of explosive compositions in storage was tried out.

On the base of the studies a Class I commercial emulsion explosive Poremit-PHA was developed for blasting operations on the open surface at the temperature from + 40°C to - 30°C in boreholes with any level of watering.

An experimental batch of Poremit-PHA was produced and its testing was carried out with the following results:
- impact sensitivity according to GOST 4545-88
    lower limit in apparatus 2   300 mm
    explosion frequency   32 %
- friction sensitivity (lower limit)   4 000 kgf/cm^2
- flash point with 60 s of time delay   228 – 232 °C
- thermal decomposition onset points   193 °C
- detonation velocity of a 45 mm charge in
    polyethylene casing   5,5*5,8 km/s

The samples of Poremit were stored in an unheated store and they were checked every month during a year. Their characteristics did not change and after 12 months of storage 45 mm cartridges of Poremit-PHA gave reliable detonation with 5.8 km/s velocity at 1.5 g/cc density.

A general conclusion is that it is feasible to develop commercial explosives, which comply with the requirements of mining industry, in articles with the diameter from 36 to 200 mm, on the base of emulsions and OST 84-1067-81 high explosives in dry or wet state (water content is up to 10 %).

UTILIZATION OF COMPOSITE PROPELLANTS WITH OBTAINING SPECIFIC PRODUCTS

A.J.SALKO, A.P. DENISJUK, YU.G. SHEPELEV
Mendeleyev University of Chemical Technology, Moscow, Russia
A.B. VOROZHTSOV
Tomsk State University, Tomsk, Russia

One of the most promising ways of elimination of large-sized propellant charges is their combustion on the bench equipment in the special pond filled with a water solution of mineral salts. The partial utilization of some combustion products in this case is possible. Thus, aluminium oxide can be produced having a wide usage dependently on its nature, structure and cleanness. Aluminium content can be up to 23% in composite propellants, that allows to consider the burning of charges with utilization of Al_2O_3 acceptable and expedient both from technical and economic points of view. The pointed purpose can be reached under condition of condensation of aluminium oxide particles being blown from a burning surface of a charge, into salt oxidizing water solutions. It will allow ensuring the best safety of a product and maintaining of ecological standards. The nature of salts, their concentration, geometrical parameters of the bench, the conditions and time of contact interaction of condensed particles with carrying medium are the factors determining the completeness of aluminium combustion and the fractional sizes of particles.

The application of special pond allows to:

- carry out the process within a compact territory;
- use minimum of technical devices (only attachment systems for propellant charges);
- ensure absence of retraction of condensed particles, their sintering and, therefore, high quality of the final product;
- guarantee rather high safety of operations;
- produce Al_2O_3 in the same pond by using charges of propellant of the various forms and weights and also crushed chips and pieces of propellant.

One of the versions of the device ensuring a high yield of product is the bench with a circuit, in the channel, of which the injection of oxidizing reactants from a lateral side is made. After passing through the circuit, combustion products go into the pond with a solution. The preliminary registration of a whole set both external and internal factors influencing the ultimate result of the implementation of such way of combustion of charges and substantiation of the specific way of the implementation of this approach are to rely on prognostic researches based on the specially constructed process model.

The mathematical model describes the evolution of the dispersed current of gas with burning particles of metal in the channel of propellant, as a result of dynamic interaction of phases in chemically active medium and changes in a fractional structure of a discrete component when the particles of different size collide. The half-closed channel of length L_0 is studied. From the part of its length L. the injection of weight and energy due to combustion takes place, and from the remaining part the liquid component is injected. The particles entering into a channel from the burning surface form a conglomerate Al_2O_3. The whole metal in a conglomerate is considered to be concentrated within a central core and in combustion of a separate particle the size of aluminium-backed core decreases accordingly to the empirical law [1]. At that, part of condensed metal oxide forms sediment on a particle, and the residue enters into a flow in form of submicron particles. The vector of forces affecting the particle in a flow is caused by forces of pressure and friction, the energy exchange between condensed and gas phases is determined by power of aerodynamic forces and heat exchange the appropriate interaction coefficients of which are described by criterion-experimental relations [2]. The particles of the same fraction are considered not to interact with each other. The Euler's model of coagulation is taken into consideration in the context of the continuous approach, and the breaking up of particles takes place in the collision with a small-sized fraction and is affected by aerodynamic forces when the Weber number reaches its critical meaning. The breaking up process is described by the model of monodispersed fragments [2].

The non-stationary heating-up and evaporation of injected liquid drops in a gas flow is described by a heat conductivity equation in spherical mobile co-ordinate system connected with the centre of the drop and movable front of phase transformation. Similar description is carried out for metal particles additionally entering into a flow when they are heated, to ignition temperature.

The set of equations expressing basic conservation laws in the integral form is written with reference to elementary monitoring volume, where the parameters of gas and condensed phases are averaged when solving. The integration is made by the method of large particles [3]. At that, those terms in right members when describe the heat exchange and force interaction between gas and particles are taken into account implicitly, whereas the coagulation processes are considered obviously. Calculation of the parameters of quasistationary current of polydispersed mixture gas - burning particles is made using step-by-step method. A the first stage, the equations of gas motion and then equations of gas and condensed phase motion disregarding coagulation and splitting of particles are integrated. At the final stage, dependently on the approximation obtained the full set of the equations is integrated.

Calculations were carried out for the channel of diameter d with the following geometrical characteristics: $d/L_0 = 0.02$, $L/L_0 = 0.1$ for the propellant with combustion temperature 3600 K at a content of the condensed phase 30%. The level of quasistationary pressure in the head part of the channel was 1 MPa. The particles entering from the combustion surface, were considered to contain 20 normal - logarithmic distributed fractions of diameters from 0.5 to 35 microns with dispersion 1.5 and mathematical expectation $d_0 = 1.25$ microns and $d_{43} = 2$ microns. It was accepted that during the combustion of aluminium 70% of producing oxides form sediment on a particle, and the remaining enters into a flow of a condensed phase with initial diameter of 0.5 microns. Agglomeration process on the burning surface was not taken into account.

On fig. 1 the change of mean-mass size of particles d_{43} along co-ordinate (x) corresponding to the length of the channel is shown. From the section $x = 0$ up to $x/L = 0.03$ the weak change d_{43} is observed, that is caused by a small difference among the speeds of particles of different fractions.

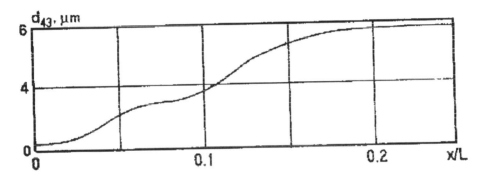

Figure 1. Change of d_{43} along the length of the channel

Then the more noticeable increase of d_{43} (owing to increasing particles speeds) is observed, and beginning from the section $x/L = 0.05$ the growth of d_{43} is again scaled down because of the burning-out of aluminium: the sizes of the particles decrease, the arrival of the condensed phase from the 0.5 microns fraction increases, the process of breaking up of the particles strengthens. Further on the flow the speed increases and coagulation processes are accelerated.

On fig. 2, the dependencies of aluminium content in a composition of the condensed phase of combustion products are shown. The curve 1 - calculation for the above described conditions, the curve 2 - for the case of injection of a cold chemically neutral reactant.

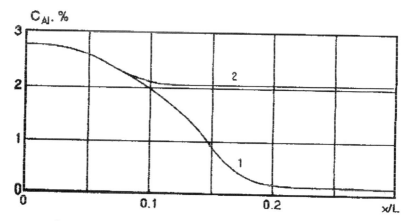

Figure 2. Change of aluminium content along the channel

Analysis of results of the calculations confirmed the functionality of the offered approach and demonstrated that the burning of the charges in chemically active medium allows to considerably increase the aluminium combustion completeness. Estimated calculations confirmed, that the efficiency of burning would be even higher for propellants with the considerable contents of pure aluminium, for which the metal agglomeration process on burning surfaces is more essential, especially at low pressures.

References

1. Pokhil P.F., Beliaev A.F., Frolov Yu.V., et al. (1972) *The combustion of powdery materials in active environments.* P.294, Nauka, Moscow..
2. Richkov A.D. (1988) *Mathematical simulation of gas-dynamic processes in channels and nozzles.*, P 222, Nauka, Novosibirsk.
3. Belotserkovsky O.M., Davidov Yu.M. (1982) *A method of large particles in gas dynamics.*, P 392, Nauka, Moscow.

CREATION OF SAFE ON MANIPULATION INDUSTRIAL EXPLOSIVES AND PRODUCTS FOR MINING INDUSTRY ON THE BASIS OF GUNPOWDER

V.P.GLINSKIY, O.F.MARDASOV, N.V.MOCHOVA, N.K.SHALYGIN
FGUP KNIIM, Krasnoarmeisk, Moscow region, Russia
B.F.OBRAZCOVA
Kalinovskiy chimzavod, Russia

Development of product based on explosive materials and gunpowder from utilised ammunition (AM) and destined for use in mining industry and for blasting jobs is made reckoning with following presumptions:

- basic consumer of developed explosive materials is the mining industry;
- resulting commercial explosives based on gunpowder and energetic materials, being withdrawn from AM must not be inferior in detonation effectiveness, technological and ecological safety in comparison with civilian industrial explosives;
- technological processes of manufacturing products of gunpowder from utilised AM must be if possible adapted to existing plants producing civilian commercial explosives;
- manufacture of developed explosives could be realised not only at industrial plants but also in arsenals and armouries where the materials are stored;
- developed explosive materials must be handed over to users for carrying out of blasting works in accordance with procedures, established by the State technical supervision of mining works RF.

Basic mass of explosive materials from utilised AM requires various levels of technological processing for their using as industrial explosives. However, it is necessary to have in view the substantiality that utilised energetic materials have high energetic characteristics and therefore also following particularities:

- high negative oxygen balance;
- they contain such components as hexogene (RDX), TNT powder etc. being more sensible as industrial energetic materials. Therefrom are resulting certain problems in the course of their application for blasting works in the mining industry;
- they need elaboration of special measures for liquidation of misfires.

The use of energetic materials resulting from dismantling of AM for blasting works in the industry can not be recommended prior to detailed refinement of safety problems.

The most feasible technical solving of safety problems is the phlegmatisation.

Taking into consideration above enumerated particularity of utilised explosive materials, the advancement of works on industrial explosive materials development on their basis was realised predominantly in following directions:

- elaboration of simplest industrial explosives (without special technological processing), representing explosive compositions of the type explosive material + phlegmatising agent – granipores, porotoles and zernite;
  In these compositions as the explosive material are used gunpowder for artillery (pyroxiline and/or ballistite types), ballistite solid rocket propellants. As phlegmatising agent are used: industrial oil, TNT, water and water solutions of oxidising agents
- elaboration of water containing explosive materials (emulsion or gel type), forming explosive compositions, undetachable part of which is water (gelpores, emulsens).

In these compositions as explosive material are used utilised gunpowders and rocket propellants and in emulsenes explosives with high disruptive force:

- elaboration of industrial explosive materials of high-filled explosives on melt basis with supplementary phlegmatisation.

In these compositions as explosive materials some pouring mixtures, containing RDX and aluminium are used. As a phlegmatising agent is obligatory used wax.

The realisation of this development direction enabled the production of industrial explosive materials – algetols.

Comparative characteristics of elaborated industrial explosives concerning the sensitiveness to mechanical actuation (shock, friction) in comparison with characteristics of gunpowder being used for manufacturing of Ammonite 6ZV as a mostly widespread product for blasting works are given in diagrams 1,2,3,4,5 (not shown here).

Comparative analysis of given data shows that industrial explosives elaborated on the basis of utilised explosive materials (gunpowder and military explosives) is comparable, with regard to sensitiveness to mechanic actuation, with civilian analogous explosive Ammonite 6ZV and is sufficiently lower as in the case of pure initial being utilised material (hexogene, pyroxiline gunpowder).

Therefore, from results of being carried out investigation follows that the elaborated assortment of industrial explosive materials technically conforms all requirements, indicated by the mining industry.

The Krasnoarmeisk Research Institute of Mechanisation drew special attention to development of processing of utilised explosives to industrial explosive materials on the basis of an emulsion. The realisation of this development direction gives wide possibilities for use of nearly all utilised in present time explosives – gunpowder and hexogene compositions – as components of explosives of emulsion type.

Compositions consisting of explosives + emulsion secures, in the first place, a sufficient manipulation and environmental security in the course of their manufacture and use. Further on it allows solving known problem of emulsion sensibilization. This problem was usually solved by means of chemical gasification or by introducing of inert particles filled with gas.

For building up of inland production of emulsion type explosive materials in cartridges, appropriate for manufacturing of charges with diameter of 45 – 120 mm, a series of industrial explosives based on the emulsion type "water-in-oil" and utilised gunpowder, for instance emulsene -P and emulsene-PB was elaborated. Here the index P stands for pyroxiline gunpowder and PB and in accordance with this stands for a mixture of pyroxiline and ballistite gunpowder.

Basic characteristics of these compositions are given in the Table 1.

TABLE 1. Basic physico-chemical and blasting characteristics of emulsene-P and emulsene-PB

No.	Characteristics	Value	
		Emulsene-P	Emulsene-BP
1	Temperature of detonation, MJ/kg	3,2	3,0 – 3,2
2	TNT equivalent	0,75	0,8
3	Volume of gaseous products, l/kg	970	1200
4	Detonation velocity in open charge $\varnothing$ 90 mm, km/s	5,2 – 5,6	5,0 – 5,4
5	Critical diameter of detonation, mm	15 - 27	80 – 85
6	Shock sensitivity (GOST 4545-88) in apparatus No.1, lower limit, mm	0 > 500	0 > 500
7	Sensitiveness to friction, lower limit, MPa	>300	>300
8	Specific volume resistivity Ω.m		90
9	Minimum ignition energy	>1	>1
10	Charge density, g/cm^3	1,50	1,45
11	Volume concentration of explosion energy, kJ/dm^3	4800	4000

Experimental unit for preparation of charges of emulsene wit diameter of 90 mm was constructed in the building No. 14 of Kalinovsky chemical works according to a project, elaborated by KNIIM:

Description of the technological process

Technological process of manufacture of emulsified gunpowder charges consists of following operations:

- acceptance and preparation of components;
- preparation of the oxidizer water solution;
- preparation of the mixture of naphtha products;
- dosing of separate components;
- preparation of the emulsion;
- forwarding the emulsion to mixing device
- mixing the emulsion with gunpowder and manufacturing of charges;
- charges numbering
- packing of charges

The phase of components preparation is completely analogized with the technological regime and machinery of existing industrial explosive materials plants. For dosing of the oxidizer solution and of the oil phase the one-screw controlled volume pumps are used. These pumps were developed in the Institute KNIIM.

Emulsifying apparatus (5) is a continuos working horizontal mixing device with rotor. The apparatus is fitted out with a jacket with inlet of cold and hot water. Separate ingredients are fed by means of controlled volume pumps under pressure; the discharge of ready emulsion is carried out by means of its forcing out by newly coming ingredients. The prepared emulsion is passing on to the collecting vessel and is pumped by means of controlled volume screw-feeder pump and is dosed to the feeding device of cartridge filling machine type ANZ-90. Here is continuously being dosed the gunpowder by help of a vibrating feeder (8).

In the course of the emulsion dosing process the checking of emulsion temperature and pressure at the dosing device output is being carried out. The manufacture of emulsified gunpowder charges is executed by automated machine type ANZ-90.

The capacity (output) of created plant is determined by the capacity of the automatic ANZ-90 and amounts up to 1800 kg/h.

Trial operation of the machinery and devices and elaboration of technological modes at the probatory unit for manufacture of emulsified gunpowder charges was carried out in the period of August – September 1997 and May – June 1998. In the course of trial operation and elaboration of the technology, the manufacture of some testing batches of emulsified gunpowder charges for preliminary testing in operating conditions of mining enterprises has been realized.

In the course of the trial operation was found that selected machinery and equipment secures realization of required regime of emulsion preparation and manufacture of emulsified gunpowder charges and may be used as a base for erection of an industrial plant.

Testing batches of Emulsene –P charges with diameter of 90 mm, prepared in the amount of 48 t. were fully conforming with the Standard TU 75 11819-90-94 and they were released for operational testing.

Preliminary testing of ZEP-90 was carried out in August – September 1998 at the testing area "Uralvzryvprom", in compliance with the program and methodic (PM 75 11819-90-94) successful.

In present time a testing batch of charges in the amount of 50 tons was manufactured. Acceptance testing of 27 tons of these charges was already carried out with positive results.

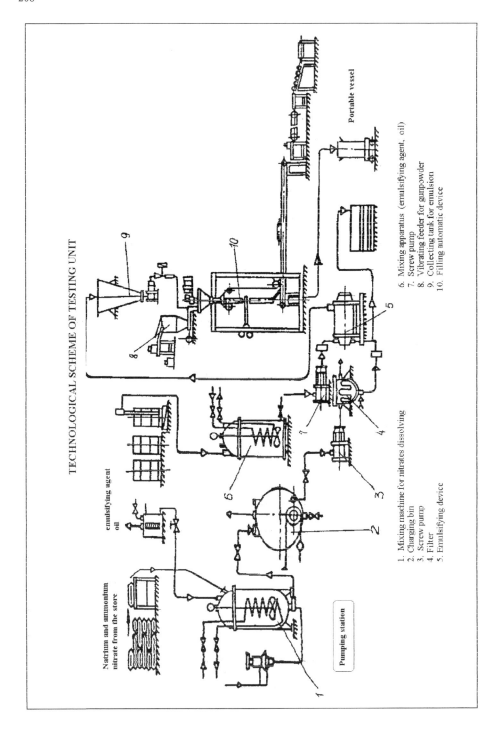

BASIC DIRECTIONS OF WORKS ON UTILISATION OF PYROTECHNIC PRODUCTS AND AMMUNITION

N.M.VARENYCH, V.G. DZHANGARIAN
FNPC NIIPCH, Russia

The Federal Scientific and Productive Centre "NII of Applied Chemistry" is a leading State Enterprise in Russia on elaboration, production and utilisation of pyrotechnic products both of military and civil assignation.

In present time in the Institute are centralised virtually all spheres of pyrotechnics. They include following works on elaboration of:

- pyrotechnic propellants for various types of rockets;
- firing devices for energetic units working on liquid and solid propellants for rocket complexes;
- passive thermal and plasmatic untrue targets;
- anti-radar and anti-laser cartridges;
- aerosol compositions for camouflage, protective and perturbing purposes and products of wide spectral actuation;
- thermobaric and firing compositions for aerial ammunition and means of close battle (head-to head);
- gas producers for acquirement of individual gases (nitrogen, oxygen);
- special facilities for combat with terrorism.

The nomenclature of already elaborated and acquired in the fabrication special pyrotechnic facilities, destined only for armament of all types of troops exceeds 450 denominations.

The concentration of scientific, experimental, constructional and technological subdivisions, manufacturing departments, testing bases and polygons in one centre, the conservation of scientific and fabrication potential and personnel allows us to carry out a full complex of works from elaboration to a mass production of pyrotechnic ammunition and products. Here belongs the production realised before and being realised in present time at factories and plants of this sphere too.

The presence of a large products nomenclature, where the mass fluctuates from grams to one and more tons, use of multiple-component mixtures and compounds, varying on chemical stability, stability, sensibility to mechanic and thermal actuation, variety of constructional schemes originated specific requirements and approaches to solving of problems of pyrotechnic ammunition and products utilisation.

Therefore, the technical-economical analysis of type pyrotechnic facilities was carried out (see Figure 1). This analysis made possible to indicate and choose the basic directions of works on utilisation of ammunition and products.

Figure 2 demonstrates these basic directions of utilisation works.

Figure 1. The nomenclature of utilised pyrotechnic products and ammunition

No.	Denomination	Amount pcs	Mass (t) of the composition	Mass (t) of the product material
1	Tracking charges of all types	9400000	180	400
2	Photo-illuminant aerial bombs	10000	400	552
	Illuminant aerial bombs			
	Firing aerial bombs	2500	150	150
	Firing cans	10100	858	1582
		5000	1030	600
3	Pyropowder engines and propellants (9D16 ed. 3D9)	7500	540	90
	"Shkval"	350	350	
4	Illuminant and signalling reactive cartridges of calibre 15, 26, 30, 40, 50 mm	15000000		
5	Smoke devices of various use and destiny	2000000		
6	Firing facilities	15000		

Figure 2

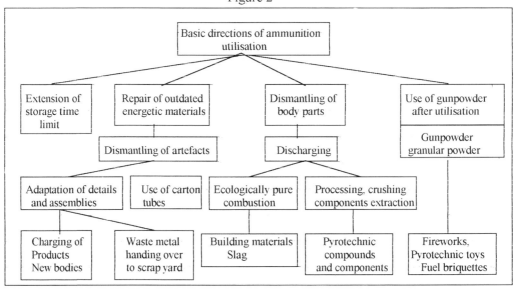

Highest economical effect produces the extension of pyrotechnic products and ammunition storage time.

The experience of application and storage of these products shows that pyrotechnic products have a great liability reserve in reliability and resources. After expiration of the storage time they may be used ad hoc.

In the course of carrying out the works on extension of guarantee terms was elaborated a new methodology, prepared necessary documentation, experimental and testing equipment.

The methodology of guarantee terms of pyrotechnic products is based on use of the method UKI, connected with execution of experimental research and method of analogues, which does not need costs for experiments.

On the basis of a demand of ordering executive of the Ministry of Defence and manufactories were in last period, in co-operation with FNPC "NIIPCh" and by use of mentioned methods, extended the guarantee terms for a series of pyrotechnic products for various applications 1,5 – 2,0 times. This concerns, for instance, the products for fireworks, for 40 and 500 mm distress- and illuminating rocket cartridges, smoke garnets and charges.

Second direction of utilisation is carrying out of a repair with exchange of some assemblies or parts or final constructional adaptation without change of target application and with maximum use of parts obtained from disassembly of outdated ammunition. This technology includes the operation of safe disassembly of body parts, adaptation of assemblies and assembly into new bodies.

In accordance with given technology were adapted 30 mm illuminating and signalling products. Reactive cartridges produced by this technology, after repair, show the quality comparable with civilian products, however they are 1,5 times cheaper.

As was found in the practice, misfiring of outdated products is caused by the part of initiation or firing. In these cases the repair consists of exchange of the initiating part, being carried out with use if safe technology.

Third direction – technology of utilising of pyrotechnic products with non-withdrawal tablets and pyrotechnic compounds.

For this class of products was elaborated the technology and equipment for liquidation of pyrotechnic elements by means of ecologically pure combustion. The combustion was executed, after dismantling of body parts, in combustion chambers (see Fig. 3), sized foe certain load according to TNT equivalent.

Condensed combustion products are settled in the rotoclone, gaseous combustion products are neutralised and deflated into the atmosphere (see Fig. 4).

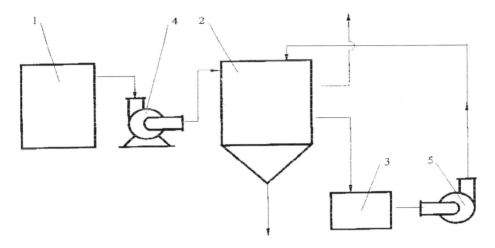

Figure 3. Technological scheme of combustion products cleaning in a rotoclone
1 – combustion chamber 4 - fan
2 – rotoclone 5 - centrifugal pump
3 – tank with absorbing solution

The slag rising after combustion of compounds is an industrial waste of third unsafeness class. This substantiality enables its add it to building materials as a filling.

Very actually is the task of increase of economic effectiveness of the utilising process by means of obtained pyrotechnic materials secondary use

Basic operations of this utilisation direction form the possibility of pyrotechnic filling withdrawal. For indivisible bodies was elaborated and erected, in co-operation with Minregion- scientific and manufacturing organisation "Hydrolaser" (town Vladimir) a unit for hydro-cutting of large bodies.

The technology of pyrotechnic powder engines (9D16, 9D 16M), pyrotechnic propellants utilisation was elaborated. This technology includes the operations of body disassembly, discharging of charges by means of pressing up or mechanical treatment with subsequent processing of pyrotechnic fillings. Further on, the obtained composition is used for disposal of biological waste and for civilian purposes.

More substantial processing makes possible the withdrawal of magnesium powder for civilian purposes, however; in this case the effectiveness of utilisation is strongly reduced.

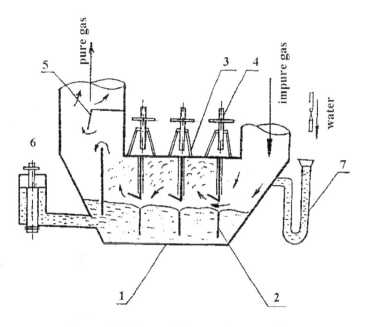

Figure 4. Scheme of the rotoclon operation

1. Casing
2. Fixed lower bridge walls
3. Movable upper walls
4. Screw jack for upper walls
5. Drop separator
6. Level control
7. Siphon for water supply

Next direction is the elaboration of products design with use of recycled in the course of ammunition utilisation gunpowder type DRP-1, DRP-2, KZDP, grained pyroxiline gunpowder and gunpowder. They may be used for manufacture of fire works of miscellaneous calibre and pyrotechnic toys.

In the course of last five years FNPC NIIPCh applies DRP, obtained after utilisation of ammunition as a push-out charge and illuminating cartridges obtained on the basis of pyroxiline powder.

Very important is in this connection the problem of liquidation of various pollution after-effects, for instance pollution with biological materials and highly poisonous chemical waste.

Death of animals, caused by various extreme conditions (earthquake, flood, and epidemic) is very dangerous for the nature, environment, animals and mankind.

Biological materials are going very quickly to be a source of accumulation of corpse poison, viruses, and bacteria and strongly deteriorate the ecological situation. They cause the contamination of ground, water reservoirs and air. The transportation of biological entities for their disposal in stationary conditions is in some extreme conditions very difficult because of absence of special transporting means. Moreover, increases the hazard of epidemic diffusion and the operation itself is very expensive. In such cases it is necessary to abolish very quickly and effectively the biological material on the place of occurrence, to assure the elimination of possible epidemic diffusion.

The problem of liquidation of biological materials, being accumulated as a result of various activities of medicine establishments, factories of food processing industry, customs-houses, markets etc. is of the same importance. Traditional processes of biological materials liquidation: burying or combusting give not any warranty of full ecological safety and need complicated and expensive equipment.

GUP FNPC "NIIPCh" elaborated, in co-operation with the Veterinary Institute VNIIVSGE RASPChN, a highly effective, ecologically safe technology of biological materials and highly poisonous waste liquidation, based on autonomous energetic sources, which does not need complicated equipment. This method is based the use of pyrotechnic powdered mixtures of filtering burning (PSFG) and was tested in the course of laboratory-stand and outdoor trials.

The components PSFG used for this process are products of solid pyrotechnic rocket propellants processing and wastes of non-ferrous metallurgy.

The selfhood of the technological process, which determines high effectiveness of PSFG in given sphere, is based on the phenomena of filtering burning.

Elaborated technology enables the solving of a complicated problem in the system of veterinary and sanitary measures on liquidation of biological materials (bodies of beasts, birds, waste of animal origin, waste of biochemical factories etc.). This is especially important, if they are infected by germs of diseases, and are dangerous for men and animals. Sometimes they threat the whole environment. This technology makes possible to burn a wide range of organic compounds – poisonous chemical waste (pesticides, herbicides, medical preparations etc – without damage of environment.

Basic technical and economical features:
Consumption of 1 kg of burned material -   0,1 - 0,2 kg of PSFG
Full combustion time -   2 – 4 hours
Burning temperature - 2000 ± 200 °C
Price of 1 kg PSFG-   3,0 – 3,5 $

POWERFUL INSENSITIVE HYBRID EXPLOSIVES USING INORGANIC PROPELLANTS/PYROTECHNICS IN CONJUNCTION WITH ORGANIC CHO COMPOUNDS FOR TAILORABLE BLAST APPLICATIONS

A.J. TULIS
IIT Research Institute, Chicago, Illinois 60616, USA

Abstract

There is no question that numerous hybrid chemical combinations of fuels, oxidizers and other energetic elements and compounds can provide much greater energy than standard molecular high explosives. Both heterogeneous and homogeneous hybrid explosive compositions are possible, and offer tremendous tailoring options, both chemical and physical, for achieving the stability, sensitivity and unique coupling of type and rate of energy release to maximize performance for specific applications. Molecular high explosives provide detonation characteristics based on factors generally limited to intrinsic stoichiometry and explosive density. In the case of hybrid explosives, both heterogeneous and homogeneous, the achievement and propagation of detonation is much more complex. A multitude of both intrinsic and extrinsic variables control how energy release is manifested and the manner and rate of such energy release. Extrinsic parameters such as particle size, stoichiometry and initiation mode, as well as intrinsic parameters such as molecular weight, density and heat of formation are involved. Furthermore, performance aspects such as homogeneity, compatibility, rheology, heat transfer, diffusion, and solubility may also be major factors. This is not a simple matter, in that so many parameters are involved and variations are possible. However, the 2, 3, or greater factor increases in energetic yields warrants their consideration as enhanced explosive compositions, especially with regard to their potential as insensitive explosives. Analytical thermochemical Chapman-Joquet characteristic computations readily demonstrate the potential substantial improvement in explosive performance of these hybrid explosives relative to the most powerful molecular explosives, although their non-ideal aspects must be considered. These non-ideal aspects, however, are often the unique features that can provide much more effective blast performance for specific applications. In this report an effort is made to provide some insight into the phenomenology and comprehension of these hybrid explosives. In addition, the conversion of propellants and pyrotechnics in conjunction with CHO compounds such as glycerine, lactose and starch to develop insensitive heterogeneous explosive compositions for blast applications is indicated. Some discussion of hybrid homogeneous explosives is also included, but experimental results of these are not included since they are not pertinent here.

1. Introduction

Major effort has been underway for decades to synthesize more powerful molecular explosives, especially attempts to achieve molecular explosive densities greater than those of RDX and HMX. However, relatively little success has been achieved in the synthesis of explosives superior to HMX, which has an explosive density of 1,91 kg/m^3 and yields 1,4 kcal/g energy in detonation; its detonation characteristics are: 9,1 km/s velocity, 38 GPa pressure, and 3100°K temperature. Although these detonation characteristics have been improved upon in recent years with the development of advanced molecular explosives; i.e., to about 2.1 kg/m3 density, 1, 7 kcal/g energy, 9.5 km/s velocity, and 42 GPa pressure (not necessarily for the same explosive); and with a few exceptions such as CL-20, most of these metastable molecules generally involve high coat and increased instability associated with their molecular complexity. Even if they could be manufactured in large quantities at a reasonable coat, their excessive sensitivities end instabilities negate their usefulness. Some novel cage molecules have also been postulated theoretically, some even synthesized in tiny quantities, but a major breakthrough here is unlikely, especially with recent military as well as industrial effort to develop insensitive explosives to replace current standard explosives, including TNT.

An alternative approach to achieve much more powerful explosives is the development of hybrid composites explosives that involve heterogeneous and/or homogeneous compositions and do not require classical chemical bonding. riot of insignificant advantage is the additional potential of "tailoring'- the sensitivities of these hybrid explosive compositions, which are generally inherently very insensitive.

The fact is that most molecular explosives are not well fuel-~-oxidizer balanced. In the case of TNT, if total oxidation were achieved, more than three times as much energy would be available as is obtained in its detonation. It has been demonstrated that the detonation of TNT in the open atmosphere provides considerably greater blast output than when detonated, underground due to blast enhancement from the oxygen availability in air. When TNT powder is dispersed and detonated in air at concentrations where oxygen from the air is substantially available, extensive blast enhancement is achieved which, considerably exceeds that achieved with similar concentrations of RDX. If dispersed and detonated in. pure oxygen or fluorine, the results would be especially spectacular. Table 1 illustrates analytical Chemical Equilibrium Code Chapman--Jouguet (CJ) results for TNT-air as a function of TNT concentration in air. The detonation of TNT releases about 1.06 kcal/g for comparative purposes.

Note that at the lowest concentration, where oxygen in the air is abundant (with carbon oxidizing to carbon dioxide), the energy output of 3,48 kcal/g is more than three times greater than the energy output at the highest concentration, where oxygen in the air is inadequate (with carbon a product as well as hydrogen and combusted carbon oxidizing only to carbon monoxide). It is not known why the energy release at the higher two concentrations is so much lower than that obtained in the release of energy in the detonation of TNT. These computations are included here only to demonstrate the potential enhancement of energy release from TNT if sufficient oxidizers were provided.

1.1 HYBRID HOMOGENNEOUS EXPLOSIVE COMPOSITIONS

In an ideal molecular explosive-auxiliary oxidizer system the oxidizer would be chemically bound to the explosive molecule either though formal chemical bonds or as adducts, i.e. molecular species bound together by largely electrostatic interactions. In such a system intimate contact between the explosive and oxidizer molecules is maintained, leading to potentially very effective and rapid chemical reactions, i.e., similar to those of molecular explosives alone wherein the chemical reaction rates are

Table 1. Computed product species and heats of reaction for the detonation of TNT in air as a function of concentration.

Species	TNT concentration in air, kg/m3			
Moles/kg TNT*	0,1	1,0	10	100
$C(s)$	trace	trace	1,67	3,13
H_2O	10,96	5,46	trace	0,04
CO	trace	23,07	28,88	26,50
CO_2	30,78	7,75	trace	0,04
H_2	trace	4,93	10,48	8,59
N_2	6,16	6,43	6,30	6,12
NO	0,44	0,18	trace	trace
H	trace	0,70	0,04	0,04
OH	trace	0,48	trace	trace
CH_4	none	none	0,04	0,18
NH_3	none	none	trace	0,04
HCN	none	none	0.88	0.92
Total moles: (gas only)	48,34	49,00	46,62	42,47
Kcal/g TNT	3,48	1,60	0,68	0,62

*Air is excluded from this tabulation except for that amount of oxygen in air consumed in the computed reaction

sufficiently fast that the mechanism of detonation is limited only by hydrodynamic responses and is independent of chemical kinetics. Hence, a hybrid homogenous explosive composition can be achieved by the dissolution of a very fuel-rich molecular explosive such as TNT in a liquid oxidizer, since the explosive and oxidizer are homogeneously mixed at the molecular level.

In an initial experiment, liquid oxygen was added to TNT powder and detonated. This was, of course, not a homogeneous composition but heterogeneous slurry. The results were a dramatic improvement in the blast performance of this hybrid explosive. Subsequently, analytical and experimental investigations were conducted involving very powerful oxidizers that would dissolve TNT without degradation, and which would result in improved stoichiometries in terms of fuel-oxidizer balance, high densities, and reasonably low haste of formation. Experiments involving the dissolution of TNT in a number of interhalogens and other powerful oxidizers, e.g., nitrogen oxides, were conducted. Atomic oxidizers such as fluorine; a gas, could not realistically considered, unless they would dissolve TNT to form a stable liquid solution at ambient temperatures. Such possibilities do exist; e.g., the dissolution of ammonium nitrate in methylamine to form an explosive liquid solution. Hybrid homogeneous explosive experiments will not be described here.

1.2 HYBRID HETEROGENEOUS EXPLOSIVE COMPOSITIONS

Organic high explosives such as TNT, RDX and PETN are chemically bonded atoms of carbon, hydrogen, nitrogen, and oxygen in a metastable molecular structure; i.e., CHNO molecules. The energy release obtained from the detonation of these molecular explosives resides in the fact that they have low or positive heats of formation, and upon detonation yield gaseous products of primarily carbon monoxide, water, carbon dioxide, and nitrogen. Hence the energy release ranges from about 1.06 kcal/g for TNT to about 1,86 kcal/g for RDX. The effectiveness of these explosives for providing tremendous blast output is due to the very high CJ pressures that are generated, attributed to the manner of energy partitioning and manifestation, with moderate CJ temperatures (about $5000°K$) but very high CJ pressures (in excess of 38 GPa for HMX); since all the detonation products are gases (about 40 mol/kg). On the other hand, pyrotechnic compositions such as aluminium (Al) and potassium perchlorate (KP) or ammonium perchlorate (AP) (AP can. also be detonated by itself and yields only gaseous species) yield condensed species with much lesser gaseous products (typically about 20 mol/kg). Furthermore most pyrotechnic compositions are generally powder mixtures of metal elemental fuels (zero heats of formation) and molecular oxidizers (considerably high heats of formation) so that, assuming they could react to CJ equilibrium fast enough, their heats of reaction (energy yields) will be diminished due to both high heats of formation of the initial molecules and the formation of condensed species as products. Nevertheless, due to the tremendous calorific potential of the product metal oxides formed, they are capable of providing energetic yields generally twice (or more) as great as those available from molecular explosives. However: the manifestation of these energy releases in detonation is heavily partitioned toward achieving extremely high temperatures (in excess of 8000°K) rather than higher CJ pressures (under 10 GPa).

Insensitive but highly energetic hybrid heterogeneous explosive compositions based on powder/slurry/other formulations of CHO, metal, and oxidizer compounds was investigated with the objective of shifting the overall manifestation of energy partitioning to reduce CJ temperatures and increase CJ pressures, mainly by increasing gaseous species yield, i.e. the CHO provides low temperatures but high gaseous species whereas the metal provides high temperatures but low gaseous species. From an energetic potential point-of-view, the following simplistic, basic chemical reaction calculations are provided:

(1) Consider a typical CHO unit (starch) molecule decomposing to C and H_2O:

$$C_6H_{10}O_5 \rightarrow 6C(s) + 5 H_2O \qquad (1)$$

The net yield is 0.53 kcal/g CHO, which can be contracted to the net yield of 1.06 kcal/g in the detonation of TNT. In fact, TIGER CJ computations that have been conducted provide CJ characteristics for this CHO that are about the same as those for AP detonation.

(2) Now suppose that Al were added to the CHO in ideal stoichiometric amount to yield aluminium oxide (Al_2O_3) as the sole product of combustion:

$$3 C_6H_{10}O_5 + 10 Al \rightarrow 5 Al_2O_3 + 18 C(s) + 15 H_2 \qquad (2)$$

The net yield is 1.83 kcal/g CHO plus Al, which is about twice that of THT in detonation.

(3) Furthermore, this does not include potential further combustion of the carbon and hydrogen products. Hence, consider the possibility of these uncombusted products mixing with air to allow complete combustion:

$$18 C(s) + 15 H_2 + 25.5 O_2 \rightarrow \sim 18 CO_2 + H_2O \qquad (3)$$

This would provide the release of additional 3.39 kcal/g of initial CHO plus Al, so that the total energy release would be 5.22 kcal/g, about five times that of TNT in detonation! Of course, the added oxygen from air would need be provided "free" here, as is the case in conventional FAE. .

The point of this discourse is to demonstrate the potential of hybrid heterogeneous explosives in (a) providing two to three times greater energies than molecular explosives and (b) allowing appropriate formulations to direct the partitioning of energies upon detonation into improved blast performance rather then excessively high temperatures. However, the heterogeneous nature of such explosive compositions has to be taken into account i.e. the actual detonation characterization of these hybrid heterogeneous explosive compositions and the multifarious parameters that will influence such.

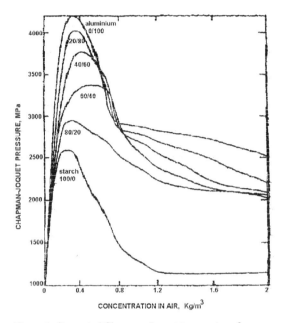

Figure 1. Computed Chapman-Joquet temperature for starch and aluminium and various mixtures thereof as a function of concentration. (Dispersions in air)

By way of example, Figures 1 through 3 are included here to illustrate the trade-off between CJ temperatures and gaseous species in the case of (starch) CHO and aluminium just discussed. These computations were conducted for dispersed considerations in air. Fig»re 1 illustrates CJ temperatures for 100 % CHO (2600°K) and 100% Al (4300°K) with intermediate mixtures falling in between. Figure 2 illustrates the corresponding CJ gaseous outputs, with 100 % Al (23 mol/kg) and 100% CHO (46 mol/kg) (neglecting the very low concentrations where CJ characteristics are poor) with intermediate mixtures falling in between. Figure 3, however, illustrates the case in point; corresponding CJ pressures for 100 % CHO (2.1 MPa) with intermediate species exceeding both values considerably, the 80/20 CHO/Al mixture in particular appearing to monotonically increase independent of the depletion of oxygen from air! Figure 4 expands the concentration range up to essentially a condensed density of 1 Mg/m3 for this 80/20 CHO/Al composition, along with TNT, in regard to CJ pressures. Note that the 80/20 CHO/Al is not substantially lower over these four orders of magnitude in concentration from a dust in air to condensed systems.

In hybrid heterogeneous powder explosive compositions, particle sizes, agglomeration, and homogeneity become paramount not only for performance, but also for the achievement of detonation. Because of the particle size limitations in typical powder candidate CHO compounds such as the polysaccharides and starches, and their tendency to agglomerate. Interest was shifted to liquid CHO compounds such as the polyhydroxy compounds, in particular glycerine. In this manner the influence of particle size and

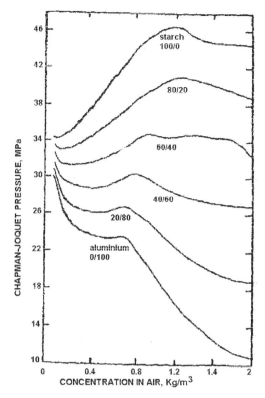

Figure 2. Computed Chapman-Joquet gas output for starch and aluminium and various mixtures thereof as a function of concentration. (Dispersions in air)

agglomeration of the CHO component, at least, was eliminated and, additionally, increased densities could be obtained. Furthermore, these formulations would no longer remain powders but could be blended and extruded, which was desired for practical applications, both military and commercial.

This still left the other two components; i.e., Al and AP powders, as major inhibiting factors for achievement /performance of detonation. Initial experiments utilized flaked Al (specific surface of 4 m^2/g) and both average 90-µm (as received) and 15- µm (ball-milled) AP. Table 2 illustrates results obtained in nominal 50-mm diameter steel tubes, With the 90-µm AP the anticipated higher densities were achieved; however, detonation failed. With the ball-milled AP stable detonations were achieved; however, upon blending these components, air was inevitably entrained so that theoretical densities could not be obtained. Special average 7 µm AP powder was subsequently procured; however, rheological difficulties persisted in attempts to blend this fine AP along with the flaked Al into the glycerine at the compositions desired. Hence, in order to achieve the theoretical higher densities that these composite explosives should provide, it was necessary to substitute 5- µm atomized Al for the flaked Al and, in most cases, to sensitize these insensitive compositions with RDX additive as well, as will be described.

Hence, in using these much coarser powders, blended compositions similar to those in Table 2 resulted in the achievement of the higher densities desired, e.g., as high as 1,7 Mg/m^3; but they failed to detonate under the same test conditions. Therefore, in order to assess the potential detonation characteristic and blast output performance in reasonably-sized tent devices, subsequent experiments with the 5- µm Al and 90- µm AP were 'Sensitized with various amounts of Class 5 RDX e.g., 10, 20, and 30 percent.

TABLE 2. Experimental detonation characterization results in 50--mm diameter steel tubes for insensitive CHO/Al/AP explosive compositions; CHO = glycerine.

Composition	Density	Detonation Velocity, m/s		Detonation Pressure, GPa	
CHO/Al/AP	Mg/m^3	analytical	experimental	analytical	experimental
50/10/40	1.07	5620	4460	9.7	8.4
50/10/40 *	1.57	7320	Failed	21.4	failed
50/15/35	1.10	5470	4260	9.5	7.8

*The AP was ball-milled in all experiments except this one.

In all cases the various compositions tested detonated, and detonation characteristics improved as the amount of RDX Was increased from 10 to 20 to 30 percent; at 30 percent it appears that optimum performance with minimum enhancement due to the presence of the very energetic RDX vas achieved. Analytical TIGER Code computations demonstrated that these small amounts of RDX do not alter the detonation characteristics to any appreciable extent; the purpose of this sensitization was an expedient to "drive" stable detonation in the composite explosives under conditions of insufficient confinement and charge diameter, not necessarily to improve detonation performance. However, the use of selective amounts of RDX, or other molecular explosive, to "tailor" the sensitivity of these otherwise excessively insensitive explosive compositions is a very viable design option.

The major obstacle to detonation of hybrid heterogeneous explosives is the particle size of the components. Even in the instances where fuel/oxidizer compounds port react sufficiently fast to propagate detonation, and the composition may be stoichiometric and very homogeneous from a macroscopic point-of-view, it is unlikely that at the molecular level

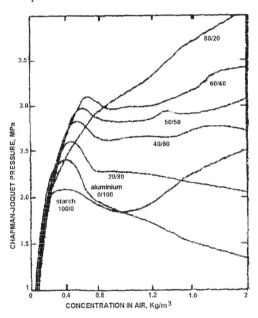

Figure 3. Computed Chapman-Joquet Pressure for starch and aluminium and various mixtures thereof as a function of concentration (Dispersions in air).

completion of reaction could be achieved within the reaction zone of ideal CJ detonation. Hence CJ equilibrium conditions are not achieved and the detonations are considered to be non-ideal.

Non-ideal detonation does not necessarily imply unstable detonation. Except for molecular explosives such as TNT, RDX, and HMX, most "enhanced" explosives such as H6 (which contains 20 percent Al) are very stable in detonation and provide exceptional performance. Non-ideality implies that CJ equilibrium conditions are not achieved due to a number of reasons, but primarily attributed to sub-critical size.

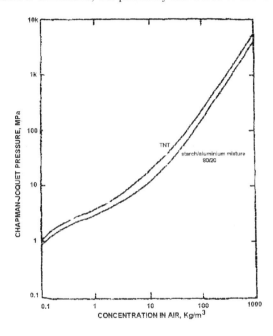

Figure 4. Computed Chapman-Joquet pressure for 80/20 starch/aluminium mixture compared to TNT, as a function of extended concentrations to condensed densities. (Dispersions in air).

Because these hybrid heterogeneous explosive compositions have extended reaction zones, as is also the case in aluminized molecular explosives such as H6, rarefactions end shock reflections can interfere within this extended detonation reaction zone. Physical transport mechanisms (diffusion and heat transfer) also may become involved. Even if that were not the case, the large particles of the components preclude efficient chemical reactions so that equilibrium is unlikely as unreacted components discharge beyond the detonation zone and may provide deflagrative "afterburn" reactions, especially if air becomes available.

Hence, subsequent experiments involved sensitization of the hybrid compositions with variable amounts of RDX additive in order to achieve stable detonations in the size/confinement test devices utilized. Experimentally, the oxidizer component was AP of both 7- and 90-μm particle size. Analytically, in addition to AP, ammonium nitrate (AN) was also evaluated, but not evaluated experimentally.

2. Analytical studies

TIGER Code (currently CHEETAH Code) CJ computations were conducted for variable compositions of CHO/Al/AP, with density as the major parameter. The CHO compound was glycerine, the metal was Al, and the oxidizer was APOD. These computations were conducted for a series of CHO: AP ratios; 0.5, 1.0, 1.5, and 2.0; for variable amounts of Al additive; i.e. 0, 5, 10, 15, 20 and 2,5 percent over the high-density range of 1 to 2 Mg/m^3. Comparable computations were also conducted with AN instead of AP as the oxidizer. Figs. 5 through 7 are included here in order to illustrate one set of these computations: i.e., for CHO: AP = 1.5. In general: (1) the CJ temperatures increase as the amount of Al increases in all cases up to (and actually beyond) 25 percent Al; (2) the CJ pressures and velocities maximize at either 0 or 5 and possibly 10 percent Al; (3) CJ temperatures are higher with AP than with AN; but (4) CJ pressures and velocities are higher with AN than with AP. Since pressures and velocities are of primary interest in detonation power, AN would appear as a better oxidizer candidate than AP.

Again, it cannot be overemphasized that these TIGER Code computations are derived from CJ equilibrium computations but that these experimental detonation results are believed to be non-ideal, although stable, detonations. Hence, equilibrium conditions at a unique CJ plane are probably not achieved and the manifestation of extended detonation reaction zones, with multiple front phenomena, are experimentally observed. Nevertheless, the major trends as illustrated in these results are as would be expected; i.e. increase of the Al content has the most significant influence on detonation temperatures due to its tremendous calorific output when oxidized, whereas because increasing the amount of Al reduces gaseous species output, decrease in detonation pressures and velocities can be anticipated. Hence, these computations do serve quite well to establish, or at least suggest, optimum stoichiometry for the most energetic compositions. In similar computations in which variable amounts of RDX were included, the amount of RDX has a very weak influence, causing a slight increase in some cases and a decrease in others, even for the same compositions but at different densities. Hence, RDX sensitization that has been necessary in conducting the reasonably sized field experiments is actually anticipated to be effective in increasing the sensitivity of these highly insensitive hybrid explosive compositions when and if needed for specific applications.

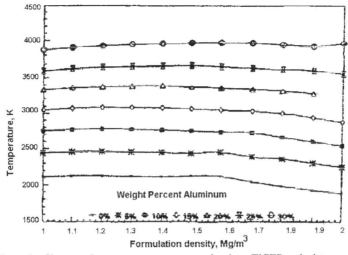

Figure 5. Chapman-Joquet temperature versus density – TIGER code data for glycerine/Al/AP explosive compositions at a glycerine:AP ratio of 1.5 and at various aluminium concentrations

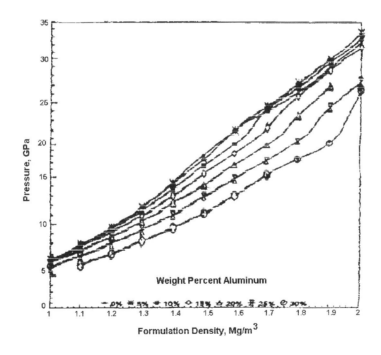

Figure 6. Chapman-Joquet pressure versus density – TIGER code data for glycerine/Al/AP explosive compositions at a glycerine:AP ratio of 1.5 and at various aluminium concentrations

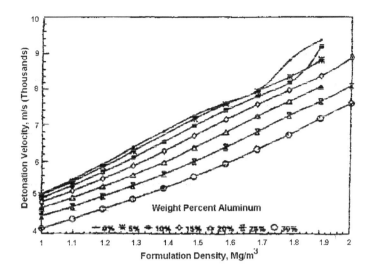

Figure 7. Chapman-Joquet velocity versus density – TIGER code data for glycerine/Al/AP explosive compositions at a glycerine:AP ratio of 1.5 and at various aluminium concentrations

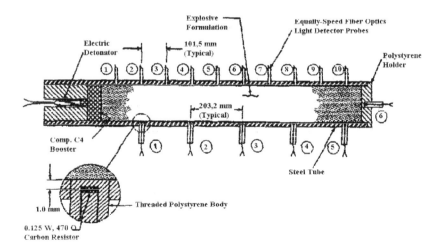

Figure 8. Schematic drawing of the instrumented confinement. Tube apparatus for measurement of the detonation velocity and pressure of explosives and explosive compositions

Regarding the environmental impact of detonation products, use of AN in lieu of AP in these hybrid heterogeneous explosives would eliminate the corrosive products due to the presence of the chlorite. For illustrative purposes, Table 3 is provided to compare the computed CJ detonation products of CHO with AP and AN oxidizers, where the CHO is glycerine, the CHO; oxidizer ratio is 1.0, the densities are 1.6 Mg/m^3, and 10 percent Al is included in the compositions. For a further comparison, the CJ detonation products for TNT are included, at the density of 1.6 Mg/m^3, as are the computed C3 detonation characteristics for all three.

The CHO/Al/AP formulations yield considerable ClH as well as small amounts of other chlorine species, including elemental chlorine. The CHO/Al/AN formulations yield detonation products much closer to those of TNT, except that the condensed species are small amounts of aluminum oxide for the former rather than the copious amounts of carbon for the latter. Furthermore, because these hybrid heterogeneous explosives can be tailored, improved stoichiometry could lead to even greater conversion to environmentally innocuous products.

3. Experimental procedures

Figure 8 illustrates the experimental detonation characterization device developed for the measurement of detonation characteristics and evaluation of detonation criteria in explosive compositions. In most experiments using these devices, nominal 25- or 50-mm diameter steel tubes are adequate and these are generally 1-m long so that there is sufficient length for detonation to stabilize and to be conclusively observed, if ft is achieved. However, in experiments involving very insensitive explosive compositions that have large critical diameters, especially when not sensitized with a molecular explosive such as RDX as previously discussed, larger diameters had to be implemented, Steel tube diameters up to nominally 75- and 100-mm were therefore used. The initiation charge was designed to create a reasonably planar detonation wave source, and generally consisted of heavy booster explosive charges of Detasheet and/or C4 of the same diameter as the inside diameter of the test device.

Larger essentially unconfined experiments were conducted upon a blast pad facility using fiber-board canister about 150 mm in diameter and 2150 mm in height. These were similarly instrumented with carbon resistor pressure gauges to monitor the detonation pressures, and shock arrival times to compute the detonation velocities. Blast pressures and impulses were obtained using piezoelectric pressure gauges mounted in the blast pad facility.

TABLE 3. Comparative analytical TIGER Code CJ detonation characteristics and detonation products for two hybrid heterogeneous explosive compositions using AP in one and AN in the other, and including similar information for TNT.

Composition	CHO/AL/AP		CHO/Al/AN		TNT	
Temperature, K	3016		2527		3712	
Pressure, GPa	20.65		19.76		20.14	
Velocity, m/s	7505		7667		7207	
CJ Species #	species*	mole %	species*	mole %	species*	mole %
	H_2O (g)	32.32	H_2O (g)	37.72	C (s)	42.42
	CO_2 (g)	19.91	CO_2 (g)	17.18	CO_2 (g)	21.67
	ClH (g)	10.70	N_2 (g)	13.50	N_2 (g)	14.53
	CH_4 (g)	9.33	CH_4 (g)	9.64	H_2O (g)	11.76
	C_2H_6 (g)	6.94	C_2H_6 (g)	6.98	CH_2O_2 (g)	2.69
	Al_2O_3 (s)	**6.03**	NH_3 (g)	6.19	CH_4 (g)	2.07
	N_2 (g)	4.64	**Al_2O_3 (s)**	**5.47**	NH_3 (g)	1.69
	CH_2O_2 (g)	3.22	CH_2O_2 (g)	2.07	H_2 (g)	1.08
	NH_3 (g)	3.18	H_2 (g)	0.84	C_2H_6 (g)	0.91
	H_2 (g)	1.56	CH_3O g)	0.36	CO (g)	0.75
	Cl (g)	0.80	CO (g)	0.03	CH_3OH (g)	0.26
	CH_3OH (g)	0.60	C_2H_4 (g)	0.01	C_2H_4 (g)	0.04
	CO (g)	0.16			CH_3 (g)	0.03
	Cl_2 (g)	0.31			CH_2O (g)	0.01271
	C_2Cl_6 (g)	0.05			CHNO (g)	0.01
Total gas, mole %		93.97		94.53		57.58

*Condensed species in bold print; s = solid and g = gas.
Species below 0.01 mole percent not included.

4. Experimental results

Results of experiments conducted in the nominal 75-mm diameter steel tubes are presented in Table 4. In all of these experiments atomized 5-μm average particle size AP was used, but both 7- and 90-μm AP were used, as indicated in the table. All these experiments were sensitized with either 15 or 30 percent RDX and had a CHO: AP ratio of about 2.0, except for the last in which it was 1.5. The detonation velocities, which are measurable very accurately, were consistently lower than the analytical values, except for the last test, which was about 1 k m/s faster than the analytical value. This composition had the lowest CHO:Al ratio of 0.8 (highest amount of Al). The detonation pressures, which are not measurable very accurately, were all in reasonable agreement with the analytical values, except for the first test, which was about 40 percent greater than the analytical value. This composition had the highest CHO:Al ratio of 4.0 (lowest amount of Al). Sufficient experiments were not conducted to attach any major significance to this. However, it does appear that the amount of RDX does not influence the performance results (analytically or experimentally) to any great extent; for that matter, neither does the particle size of the APOD. This is not unexpected, since the role of the RDX additive is to sensitize these compositions to propagate detonation and not to improve the performance, as discussed earlier.

Experiments conducted in the nominal 100-mm diameter steel tubes are presented in Table 5. In all three of these experiments the CHO was glycerine, and the CHO:AP ratio was 2 (the ratios in the table are rounded off). The mean particle sizes of the constituents were 5-pm atomized A1, 7-pm AP, and 22-μm Class 5 RDX. A commercial blender was used in order to achieve reasonably homogeneous compositions.

The first experiment in Table 5 is the identical composition tested in the nominal 75-mm diameter steel tube included in Table 4. The detonation velocity in the larger, nominal 100-mm diameter steel tube was about 100 m/s greater, but still more than 1 km/s less than the analytical CJ value. The detonation pressure was essentially the same. The experimental variance of measured detonation pressures using the carbon-resistor pressure gauges is 10 to possibly 20 percent, whereas the detonation velocity measurements are quite accurate.

TABLE 4. Experimental detonation characterization results in nominal 75 mm diameter steel tubes for hybrid CHO:Al explosive compositions senzitized with variable amounts of RDX; CHO = glycerine.

Composition ROX/CHO/Al/AP	Density Mg/m^3	Detonation Velocity, m/s		Detonation Pressure, GPa	
		analytical	experimental	analytical	experimental
30/40/10/20.	1.47	6868	6480	16.15	22.36
29/36/19/16#	1,58	6721	4235	16.47	14.43
3013340/17	1,64	6920	6178	17.87	18.28
15/43/20/22#	1.57	6637	5275	16.78	12.16
15/43/20/22	1.52	6429	5156	17.87	18.28
30/24/30/16	1.70	6195	7154	14.70	11.79

*The AP particle size was 7 μm in these experiments, 90 μm in all others.

In the second experiment in Table 5, an effort was made to assess the influence of the amount of A1, Therefore the A1 was increased from 20 to 30 percent, while the CHOW ratio was maintained the same and the amount of RDX sensitizer also remained the same. The results provided no significant change in either detonation velocity or pressure.

TABLE 5. Experimental detonation characterization results in nominal 100-mm diameter steel tubes for hybrid CHO/Al/AP explosive compositions sensitized with variable amounts of RDX, CHO = glycerine.

Composition ROX/CHO/Al/AP	Density Mg/m^3	Detonation Velocity, m/s		Detonation Pressure, GPa	
		analytical	experimental	analytical	experimental
15/43/20/22	1.58	6637	5360	16.78	11.92
15/37/30/18	1.58	6173	5400	14.94	10.56
30/27/30/13	1.76	6422	6531	15.89	18.03

In the third experiment in Table 5, the amount of RDX sensitizer was increased up to 30 percent to establish whether the otherwise comparable second experiment in Table 5 was not adequately "driven" by the RDX sensitizer. In other words, the CHO:AP ratio remained the same, as did the amount of Al, but the amount of RDX was increased from 15 to 30 percent. The results indicated that the amount of FORDX did indeed enhance detonation performance extensively; in fact, to essentially the analytically computed CJ detonation values. This experiment was the only one that yielded both pressure and velocity characteristics that exceeded the CJ values.

Because of the CJ-type results in this third experiment, an experiment was conducted with this same formulation in the nominal 150-mm diameter fiberboard canister and detonated upon the blast pad facility in order to obtain blast pressures and impulses as functions of scaled distances. The results of this experiment are presented in Figs. 9 and 10 for pressures and scaled impulses, respectively. Data is also included in these figures for a cast TNT hemisphere charge for comparative purposes. As can be seen, the output in both terms of blast pressure and impulse is as good or superior to that of the cast TNT.

5. Conclusions

It is concluded that these insensitive hybrid heterogeneous explosive compositions, based on the use of organic CHO compounds, in particular as liquids, in conjunction with highly energetic pyrotechnics such as Al and AP or AN, have the potential to evolve into a new category of insensitive, highly energetic and possibly more environmentally acceptable explosives, with the probable additional advantage of very moderate cost. The enhancement of detonability by the incorporation of appropriate amounts of molecular explosives such as RDX, has tremendous potential for tailoring various these insensitive hybrid explosive compositions to desirable sensitivity levels as well.

Although earlier studies demonstrated that these explosive compositions are capable of detonation directly if sufficiently small pyrotechnic particle sizes are used, blending and

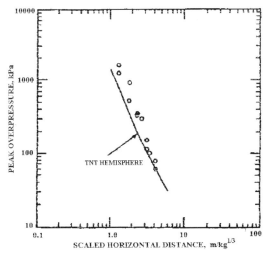

Figure 9. Blast peak overpressure from detonation of 7.5 kg 30/26.7/30/13.3 RDX/Glycerine/Al/AP

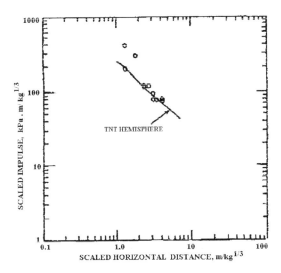

Figure 10. Blast scaled impulse from detonation of 7.5 kg 30/26.7/30/13.3 RDX/Glycerine/Al/AP

extrusion problems with these very fine powders will require appropriate equipment for blending and extrusion. Such equipment is available and would allow the formulation of these highly insensitive hybrid explosive compositions to achieve the adequately high densities needed for effective detonation characteristics and blast performance.

The use of molecular explosive sensitizers may be used to overcome problems of using inadequately small particle size constituents, as illustrated in this work, and may be useful for effective tailoring of detonation characteristics as well. However, such molecular explosive sensitizer would not be needed if appropriate and/or optimum particle sizes were effectively used.

The more environmentally acceptable oxidizer AN in place of the AP is anticipated to provide as good or superior performance. Other energetic oxidizers are also expected to eventually be investigated. Because of the tremendous tailorability that is conceivable with these extrudable explosive compositions, especially by the sensitization capability achievable with very moderate amounts of RDX or other explosive, both sensitivity and performance should be readily tailored for specific, and quite diverse, explosive blast applications.

It should be noted that the glycerine is an integral component of these hybrid explosive compositions and also provides extrudability potential. However, glycerine may not be the optimum CHO component; a denser liquid with higher oxygen content and perhaps lesser heat of formation would provide even better performance.

CONDITIONS OF AMMUNITION UTILISATION IN PEOPLE'S REPUBLIC OF CHINA

TSIN CHANJUNG
NORINCO, Beijing, China

Ladies and gentleman,

First of all I should like to express my warm-hearted thanks to the Administrative Committee of this Workshop for the invitation and to the presidency for the possibility to inform you shortly on conditions of ammunition utilisation in our Republic. I am first time taking part in such conference. I am, however, familiar with the problems of ammunition and gunpowder utilisation for a long time. Our corporation includes 19 enterprises and one Research Institute, and is specialised in the field of production and sales research of military industrial explosives and rocket propellants. The Corporation was re-created from the Chinese Department of Gunpowder and Energetic Materials. It is in present time the sole control organisation in China on military and industrial gunpowder, explosives and rocket propellants. Our Corporation produces exclusively TNT, hexogene (RDX), nitro-glycerine etc. In our production program are included one- two- and tribased propellants, and explosives, gunpowder for hunters and sportsmen, propellants for climatic rockets, emulsified and gelatinised explosive materials. Further on detonation seismic charges, detonators and detonating cords, pyrotechnic products etc. From this follows that the Corporation fulfils universal functions, from scientific research to production and sales control. As a sole in our country organisation, being responsible for utilisation of gunpowder and ammunition, we pay serious attention to meetings and conferences as present workshop. In present time we have wide connections and contacts with large world organisations and companies, among them with Russian centres and institutes, operating in this field. The target of my presence here, at given symposium, is to meet colleges and get acquainted with their experience. Even though we are, in the case of present meeting, in the position of new-invited, we desire to join this across-the-board process, because the problem of utilisation of gunpowder and explosive materials is relevant for us too.

Further on I should like to give you some information on the present state of this thematic in China.

Chinese government pays a very intensive attention to this process and releases early certain financial resources for mastering of utilisation problems. In proximate solving of these problems are engaged the State Planing Committee and the Committee of Resources.

Utilised ammunition and explosive materials are classified into two parts:
1) Getting physically old ammunition, as cartridges, bombs, mines, torpedoes etc.
2) Long time stored and outdated gunpowder and explosive materials.

General principles and methods of their processing:

Burning and liquidation of old ammunition and explosive materials by mass burning are strictly forbidden because the problem of environmental safety is now actual as never before.

In the process of utilisation of one-based explosives many readjusted product are obtained. For instance, by use of recrystallization more purified hexogenes are produced.

The recovery of a part of ammunition is carried out in special works. Here the gunpowder is utilised to industrial explosive material.

In present time many of our works produce emulsified and gelated explosives manufactured from one-based types of gunpowder. Comparatively lately we started the utilisation of double-based gunpowder and produce gelated explosive materials. Their dosing for various charge types amounts from 10 % to 50 vol. %. The density of explosives is not higher than $1,0 - 1,5$, the detonation velocity makes $4300 - 6000$ m/s. Storage time amounts one year.

More complicated problem, which we meet in the last time is breaking-up of large dimensional charges of gunpowder and propellants. This problem was discussed here. We shall very seriously investigate the opinion of our colleges.

Some perspective ideas for the future:

For activation of the utilisation process, in addition to the increasing of subsidies it is inevitable to expand the ranges of utilised materials application.

In China work some hundreds of factories producing industrial explosive materials. We are intending to offer to the government the application of utilised explosives and propellants in the form of industrial explosives and for securing the safety to establish specialised utilisation centres in separate regions. These centres would be obliged to solve the problems of transportation and storage of energetic materials.

I was pleased to have the possibility to inform you shortly on the state of gunpowder utilisation in China, on some principles of our work and some plans for the future. I hope gratefully that present here colleges will understand our position. We are ready to proceed in the future our contacts and co-operation. The science has no borderlines and for welfare of the mankind the West and East must be unanimous.

We suggest to organise such meetings more frequently and we hope that present administrative committee will be transformed to the World Company of gunpowder and explosives utilisation as a company of pyrotechnics and nitroexplosion. The financing may be based on members contributions and/or sponsoring. It seems to be beneficial to carry regularly out meetings, where one or two thematic will be discussed more comprehensively. We all know that after faster renovation of ammunition will follow more sharp and difficult increase of problems of gunpowder and explosives utilisation. This process has no borders and therefore we must not fear to be threatened by unemployment.